WHAT EVERY ENGINEER SHOULD KNOW
A Series

Editor

William H. Middendorf

Department of Electrical and Computer Engineering
University of Cincinnati
Cincinnati, Ohio

What Every Engineer Should Know About

Material and Component Failure, Failure Analysis, and Litigation

Other volumes in preparation

What Every Engineer Should Know About

Material and Component Failure, Failure Analysis, and Litigation

Lawrence E. Murr

Oregon Graduate Center
Beaverton, Oregon

MARCEL DEKKER, INC.   New York and Basel

Library of Congress Cataloging-in-Publication Data

Murr, Lawrence Eugene.
    What every engineer should know about material and
component failure, failure analysis, and litigation.

    (What every engineer should know ; vol. 20)
    Bibliography: p.
    Includes index.
    1. Fracture mechanics. 2. Materials. 3. Products
liability. I. Title. II. Title: Material and component
failure, failure analysis, and litigation. III. Series:
What every engineer should know ; v. 20)
TA409.M86   1986   620.1'12   86-24058
ISBN 0-8247-7732-8

MARCEL DEKKER, INC.
270 Madison Avenue, New York, New York 10016

Current printing (last digit):
10 9 8 7 6 5 4 3 2 1

PRINTED IN THE UNITED STATES OF AMERICA

Affectionately dedicated to my granddaughter

JANET ELIZABETH COLORADO

About the Author

LAWRENCE E. MURR is Director of the Office of Academic and Research Programs and Professor of Materials Science and Engineering at the Oregon Graduate Center in Beaverton. The author of over 350 journal articles as well as a consultant, editor, and reviewer, Dr. Murr has published 13 books, including *Metallurgical Applications of Shock-Wave and High-Strain-Rate Phenomena, Industrial Materials Science and Engineering, Electron and Ion Microscopy and Microanalysis,* and *Solid-State Electronics* (all titles, Marcel Dekker, Inc.). He is a Fellow of the American Society for Metals and a member of numerous other professional societies. Dr. Murr received the B.Sc. degree from Albright College, B.S.E.E. in electronics, M.S. in engineering mechanics, and Ph.D. in solid-state science from Pennsylvania State University.

Preface

Materials engineering is a broad field involving not only the design and fabrication of materials, but also their efficient application in structural designs, product fabrication, and related industrial technologies. A crucial feature in materials fabrication involves an understanding of elemental chemistry and crystal structures, while applications normally hinge upon the relationship of those structures to properties and behavior of materials.

This book is intended to provide a fundamental overview of the principles of materials science and engineering, including the role of crystal structures and crystal imperfections in determining and controlling the properties of materials. Principal methods of examining the fundamental nature of materials, of testing materials (especially in the context of elucidating the mechanical properties and behavior of materials), and examining the failure and failure modes of materials, particularly metals, are presented using specific case examples to illustrate both the methods and the principles. Nine such case histories are integrated into this book as summaries of the litigation process.

It is important for engineering students, engineering technology students, and even practicing professionals, including attorneys (particularly those involved in aspects of product liability litigation or in litigating cases involving failure or suspected failure of materials), to have some basic notions about materials, fundamental aspects of their failure, and methods for investigating those failures. Learning something about contemporary materials and how they fail may also contribute to the prevention of product liabilities connected with poor choices of materials in product development, or of particular uses of materials in manufacturing processes.

Lawrence E. Murr

Contents

What Every Engineer Should Know About

Material and Component Failure, Failure Analysis, and Litigation

1

Introduction

An appreciation for the science and engineering of materials and their properties in relation to how they might fail will provide a framework for decision making and forensic analysis by an engineer or an attorney. These are crucial features of litigation involving the determination of liability in product or materials failure.

An attorney once came to me, seeking my advice in litigating a claim in which a client had severely injured an eye while working on a large grinding machine. The claim sought compensation from the company on the grounds the operation had engineering and design flaws that resulted in improper safety features. Resolution of the claim hinged on being able to show that a metal fragment taken from the worker's eye did indeed come from the machine.

I advised the attorney that it would be a relatively simple matter to place the metal fragment recovered from the worker's eye in a scanning electron microscope fitted with an energy-dispersive X-ray spectrometer, so that a spectrum of its elemental (chemical) composition could be displayed on a cathode-ray tube. In addition, by extracting a similar, fine fragment from the suspect machine part, a similar elemental spectrum could be simultaneously displayed under identical operating conditions and the double spectra photographed as unambiguous evidence.

Unfortunately, the original eye fragment had been given to a chemical laboratory, where it was reacted in acid to dissolve the elements for a liquid-phase chemical analysis. The analysis itself was not successful and the evidence destroyed in the process. Consequently the claim was lost.

The message in this example, if there indeed is a message, is that neither the attorney nor the chemical laboratory had any notion of modern analytic or microchemical instrumentation, and they were not familiar with modern failure analysis tools or methods of nondestructive evaluation and analysis. Moreover this was a metal failure, and neither a metallurgical nor a materials specialist was consulted when it might have made a difference in the outcome of the litigation.

Liabilities as prescribed in liability laws vary throughout the United States but are more often than not dependent upon the demonstration of fundamental design or engineering flaws in the case of failures or suspected failures in materials systems. To understand materials failures requires a fundamental knowledge of materials properties and behavior within the broad discipline of metallurgical and materials sciences and engineering. Materials systems range over a broad spectrum, including metals, alloys, and ceramics (and glasses), semiconductors, polymers, and plastics, and combinations of these materials — composites, such as graphite fiber-reinforced plastics, oxide particle-reinforced metals, and multilayer microelectronic device structures. Products of industry can involve many combinations of these materials and materials systems, and product failure may involve failure of only a single component of a more complex materials system. Failure includes catastrophic fracture under stress, corrosion, delamination, and more complicated interactions, which could involve stress, temperature, and other environmental factors, or combinations of variables.

This book is primarily concerned with fracture and catastrophic failure of materials, primarily products characterized mainly as composed or as fabricated of metals and alloys. Fundamental studies of fracture (as fractography) are presented. In defining fracture modes in metals, alloys, and other materials, some fundamental properties of these metals and materials are briefly reviewed from the perspective of establishing some sensitivity to materials science and engineering. After defining fracture modes, the analysis and characterization of fracture and fracture surfaces are described in the context of analytic techniques and instrumentation for viewing fracture surfaces and for probing the structure and chemistry of fracture surfaces. The use of electron and ion beams to probe and view surfaces by electron and ion microscopy and spectroscopy is emphasized in discussions of analytic instrumentation. This also involves discussions of the advantages and limitations of such instrumentation. Finally, this is interspersed with and followed by examples of applications of these principles and techniques in litigation, generally involving the determination of liability or fault in product or materials failure. Some of these examples are extensive case histories, with excerpts from depositions, preparation and organization of evidence, and trial outcomes or other resolutions of claims.

The organizational structure and implicit philosophy of presentation is intended to create an appreciation for the science and engineering of materials. More specifically, this presentation is intended to create an appreciation for the properties of materials in connection with failure, a recognition of the uniqueness of failure modes, especially in metals and alloys, and an awareness of the modern methods available for characterizing and identifying failures through an examination of fracture surfaces and the structure or chemistry of materials. It is hoped that this book will provide a framework for decision making, analysis, and development of supporting conclusions when an attorney has to deal with a broken product or component and determine the conditions under which it failed, whether the failure was due to faults in the material or errors in its use through the faulty design or engineering of a product.

2

Fundamentals of Materials Systems: Some Elements of Materials Science and Engineering

Structure, including atomic structure, must play a fundamental role in the behavior of materials.

Before we discuss materials analysis and characterization, particularly materials failure and materials systems failure, it is useful to review some of the fundamental aspects of materials science and engineering. It is also logical to begin with some elements of materials properties and behavior since these features characterize the failure modes of materials. Obviously, it is not practical to review the whole of materials science, even in the context of elementary principles, so this review of necessity is brief and largely graphic. The reader will need to consult some of the key literature (provided in the Notes) for a more detailed description of specific issues and areas.

There are numerous classes of materials grouped according to structure, properties, and behavior. Figure 1 is a pictorial classification for the more prominent engineering materials in use today, exclusive of wood, which is not discussed in this review. The metals group has been left unshaded to indicate not only a prominence of metal applications, but also the preference given to metallic materials in this chapter. This pictorial classification is intended to show the mixing of these materials to form composites and materials systems. A simple example of a materials system could involve a spark plug, consisting of several different metals and a ceramic, or a computer chip, consisting of various semiconductors, ceramic insulating components, metal conducting layers, and plastic encapsulations.

In any of these general classes of materials, there are two fundamental arrangements of the atoms and/or molecules: orderly arrangements, which are called crystals or crystalline arrangements, and structureless or random arrays of the atoms or molecules. Materials with orderly arrangements or collections of orderly arrangements are said to be crystalline materials; those with no organization or structure are said to be amorphous materials. The behavior or properties of materials is often very much dependent upon this crystalline or amorphous structure or upon the degree to which these structures might be intermixed in a material or a materials system. Figure 2 illustrates these fundamental differences in the structure of materials.

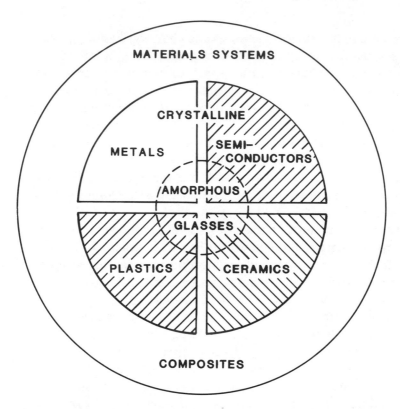

Figure 1. Simple pictorial classification of materials and materials systems. The dotted circle is indicative of the fact that any of the four major classes can be regarded as glasses or as having glassy structures under certain conditions. Composites can be developed by a variety of combinations of ceramics in metals, ceramics in polymers, and so on.

It is certainly apparent even from the simple schematic illustrations in Figure 2a and b that structure, even atomic structure, must play a fundamental role in the behavior of materials. For example, consider conduction of electricity, represented by the propagation of an electron from one side of the arrangement to the other in Figure 2a or b. The electron, with all other conditions identical, would less readily traverse the structure in Fig-

ure 2b in a given time in the same way it would take a runner
more time to run the same distance through a randomly planted
cornfield than simply to run down a row in a regularly planted
cornfield. The same is generally true for demonstrating other
differences in properties or behavior.

It is helpful and sometimes important to know something
about the atomic structure of a material, but a general awareness
of the behavior of generic groupings of materials implicit in Figure 1 is sufficient in approaching certain problems involving
liabilities. For example, most metals are ductile and can be bent,
but glass is generally brittle and will not bend. Plastics are ductile
but not as strong as metals, generally, but when they are cold
(say, close to the freezing point of water), they can become
brittle. Obviously, there are fundamental reasons for these differences. Most metals and generic glasses, for example, are characterized atomically and molecularly as shown in Figure 2a and b,
respectively.[1] The following case example serves to illustrate this
point.

Case Example 1

In the late 1960s in a Los Angeles District Court, a young
black man, about 17 years of age at the time, was on trial
for attempted vehicular homicide. It was claimed that this
youth had driven through a police barricade in a deliberate
attempt to run down police officers. The youth claimed that
he had turned into the barricaded street on his way to his
home a few blocks away and had been shot at immediately
by officers, their gunfire hitting him and causing him to lose
control of the vehicle. The prosecution claimed that the
youth was not fired upon from the front but that the police
fired from the rear in an effort to stop him after he had
crashed through the barricade. Consequently, the police fired
in self-defense in an effort to stop the vehicle.

During the trial, ballistics experts had sought to demonstrate that the bullet holes in the vehicle, especially those in
the windows, all supported the contention that the police
fired from the rear or toward the rear of the vehicle. Detailed

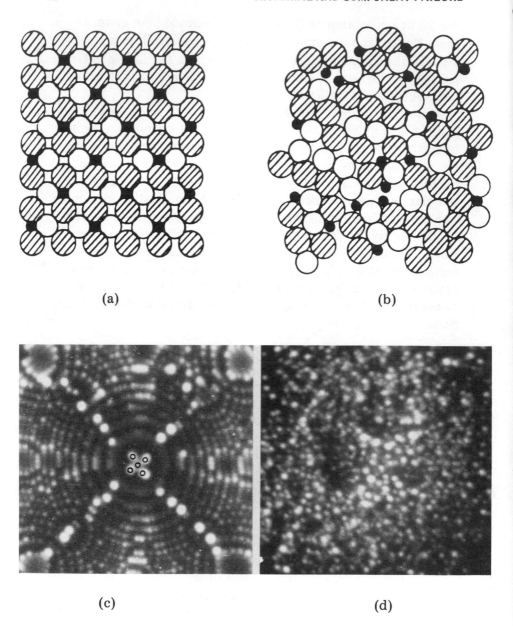

(a)

(b)

(c)

(d)

Figure 2.

accounts were presented to illustrate that the glass fractured as it should have for bullets entering from the rear. Obviously it was a simple matter to show that holes in the metal structures, such as rear fenders and bumpers, were due to bullets fired from the rear because of the way the metal was stretched and torn in the penetration process. But glass is brittle and does not easily demonstrate these unique features.

The public defender asked me to examine the evidence and circumstances in the case because of my research and interest in shock and ballistic phenomena, especially the response of metals to shock deformation. The case hinged upon demonstrating that at least one of the bullets had entered the front of the vehicle and had nearly severed the youth's right arm at the shoulder. On examining the vehicle, which contained more than 100 bullet holes from high-powered rifles, I noticed 14 bullet holes in the windshield. This was a 1960s-vintage automobile with safety glass of that period, called triplex glass. Unlike modern safety glass, which shatters into very small fragments on sufficiently violent

---

Figure 2. Examples of crystalline and amorphous structure of materials. (a) Regular arrangement of atoms characteristic of crystalline material. (b) Random arrangement of the same atoms characteristic of amorphous material. (c) Field-ion microscope image of regular atomic structure of iridium metal whisker. (d) Field-ion microscope image of random (amorphous) atomic structure of an iron-boron alloy whisker. (See L. E. Murr, *Electron and Ion Microscopy and Microanalysis: Principles and Applications*, Marcel Dekker, Inc., New York 1982, for a detailed explanation of the principles and operating features of the field-ion microscope. This microscope allows the atomic arrangement of the atoms in the surface of certain metallic whiskers to be viewed directly. Indeed, the magnification of the images in c and d is several million times).

impact, triplex glass contained an inner plastic layer that prevented the glass from shattering. It also prevented passengers
from flying through the windshield, however, and for certain
impact conditions was definitely not safe. However, since the
windshield was a lamination of glass and plastic, it simply
would not respond like window plate glass to a penetrating
bullet. It was reasoned that the high-velocity bullet would
stretch the plastic interlayer by frictional and impact or
shock heating when it penetrated, and by the stretching or
stressing associated with penetration, as shown schematically
in Figure 3.

Figure 3 assumes that the plastic effectively would flow
in the direction of penetration and remain in that direction
after penetration as a permanent indication of a directionally

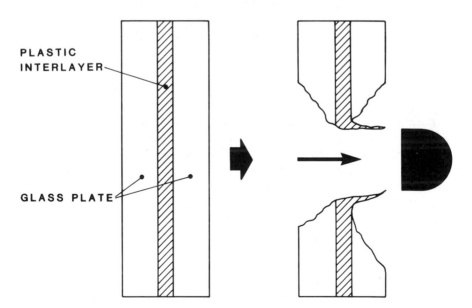

Figure 3.   Schematic view of older vintage triplex automobile
safety glass in cross section showing the directional stretching of
the inner plastic layer during penetration by a large-caliber bullet.

specific process. The task then was to illustrate unambiguously that at least one of the bullet holes in the vehicle's windshield corresponded to the situation shown schematically in Figure 3. Three bullet holes in the driver's side of the windshield were indeed observed to prominently show the plastic flow geometry indicative of front-entry penetration, and on the basis of this evidence the youth was acquitted.

Figure 3 is also an illustration of a composite or a materials system. The significant feature in such systems, especially in Figure 3, is that the materials components behave differently, have different properties, and are structurally and chemically different. Each component therefore responds differently to a deforming force or related environmental factor.

Metals are usually thought of in generic terms, although pure metals are single elements, such as gold, silver, or aluminum, and alloys are systems of elements or homogeneous mixtures of metals, such as zinc and copper to form brass or iron, chromium, and nickel to form stainless steel. Metals and alloys have different crystalline structures, but in industrially and commercially common metals and alloys (which are crystalline), most metals and alloys have one of three principal crystals: body-centered cubic (bcc), face-centered cubic (fcc), or hexagonal close-packed (hcp). These are only three of the fourteen different fundamental crystalline structures illustrated schematically in Figure 4 along with some common metals and alloys with these crystalline structures and their elemental compositions. Figure 4 also illustrates the crystalline structural transformations that can occur in other materials in response to certain combinations of heat and pressure and of chemistry. There are of course other more subtle features that allow, or provoke, structural transformations. For example, cobalt transforms from fcc to hcp under certain conditions. Combinations of elements or metals to form alloys also develop different crystalline structures. Notice that both nickel (Ni) and aluminum (Al) as pure metals possess a face-centered cubic (fcc) crystal structure. When Ni and Al are combined in equal portions to form the alloy NiAl, the crystal structure is

Face-centered cubic

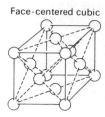

Body-centered cubic

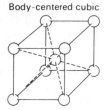

Hexagonal

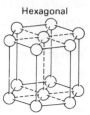

aluminum (Al)   iron (α-Fe)   cadmium (Cd)
nickel (Ni)   chromium (Cr)   titanium (Ti)
lead (Pb)   tungsten (W)   zinc (Zn)
α-brass (70% Cu, 30% Zn)   β-brass (60% Cu, 40% Zn)   α-titanium alloys
stainless steel (73% Fe,   iron-nickel (90% Fe,   magnesium-zinc alloys
   18% Cr, 9% Ni)   10% Ni)   (92% Ti, 8% Al)

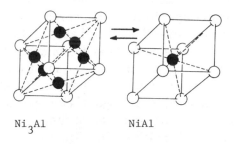

Ni$_3$Al   NiAl

Simple Structural and Chemical
Transformations

Figure 4. Common unit cells representing fundamental crystal
structures in metals and alloys and some corresponding examples.
The bcc to fcc transformation shown can also occur for a variety
of the metals and alloys. The *a*-Fe (bcc) to *γ*-Fe (fcc) transformation is another common example.

body-centered cubic (bcc). Nickel and aluminum can form another
alloy with different composition, however, namely, Ni$_3$Al, and
the logical crystal structure that does in fact occur is face-centered
cubic (fcc).

ctural (or microstructural) transformations are very
materials and in materials systems because their
d behavior are often dependent upon a particular
ructure, and when the structure changes, behavior
rns out, for example, that many alloys with body-
ic crystalline structures, although strong, tend to be
ersely, most alloys with a face-centered cubic crystal-
re tend to be more ductile. So a transformation in
cture can lead to a local or an overall embrittlement of
se ductile structure. Temperature can also have a
ffect on this ductile-brittle transition, and bcc metals
which may be ductile at elevated temperature, can
ittle at lower temperatures.

symmetrical or orderly arrangement of different atoms
such as the Ni-Al alloys shown schematically in Figure 4,
ed to as *ordered alloys*, as opposed to a random arrange-
the specific atoms even in the unit cell arrangements
Obviously this order-disorder phenomenon can also be
ed another kind of transformation, and it will depend
e specific composition of the alloy, or stochiometry. For
le, if the nickel content of the Ni_3Al alloy were increased,
ructure might become disordered and the nickel atoms begin
ubstitute for some of the corner sites in the fcc unit cell
wn. This might occur at random corners, and the alloy would
ereby become disordered as the unit cells were stacked together.

It is important to realize that the crystalline units shown in
Figure 4 are referred to as unit cells. These represent the smallest
geometrically distinct arrangements of the particular crystalline
structure. The dimensions of a unit cell can be determined experi-
mentally using electron or X-ray diffraction,[2] and large, practical
crystals of a material are seen to be composed of enormous numbers of such unit cells stacked in perfect periodic arrangements.

When metals and alloys solidify from the melt they normally
form crystals with one of the crystal structures shown in Figure 4,
but a solidified ingot of such metals or alloys is not normally a
single crystal. Very special solidification processing is required to
produce enormous single crystals, and this is indeed the case in

Figure 5. Optical metallographic view of etched grain or crystal structure in a commercial (304) stainless steel. Note the relative size of these grains or crystallites indicated by the magnification or scale marker.

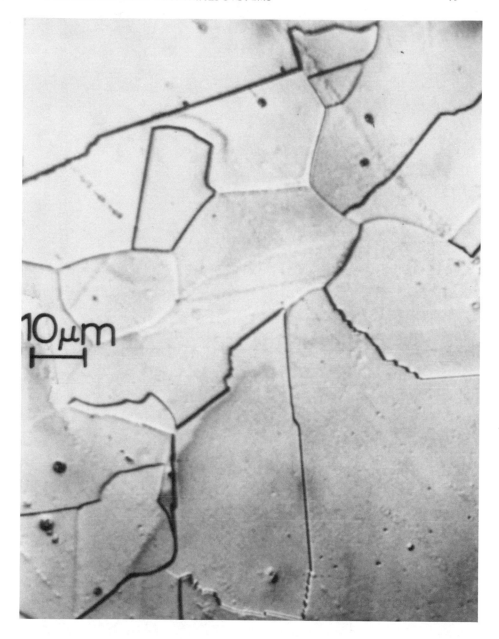

Figure 5. (Continued).

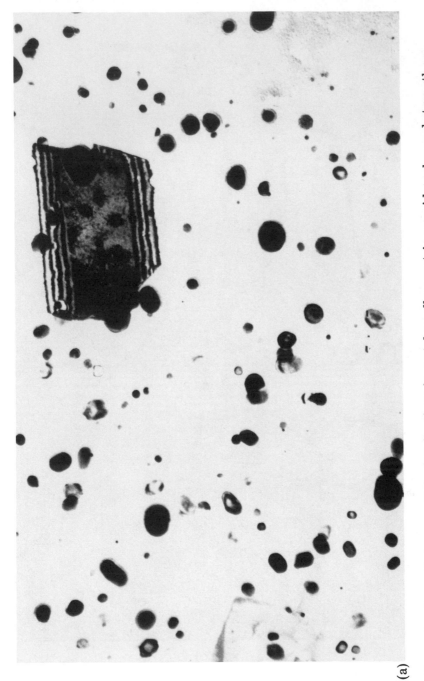

(a)

Figure 6.   Second-phase particle distributions in metal or alloy matrix provide enhanced strength or hardening. (a) Thorium dioxide particles dispersed in nickel. A small occluded twin grain is also shown.

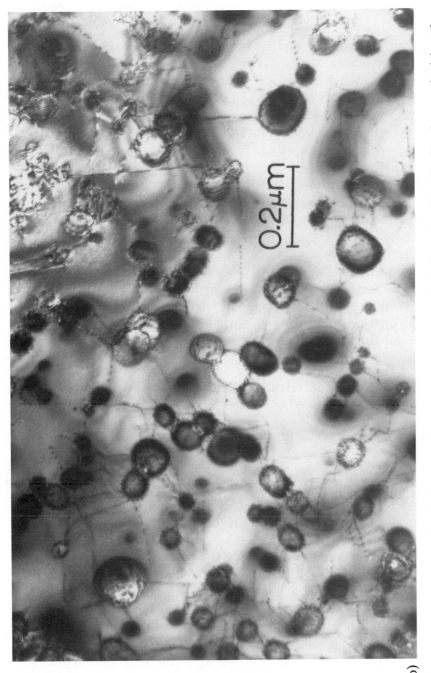

0.2μm

(b)

Figure 6. (Continued) (b) NiAl precipitates in an iron-based alloy (Fe + Cr, Ni, Al). The images in (a) and (b) are transmission electron micrographs obtained by preparing thin electron transparent films of the materials and viewing them in the electron microscope at an accelerating voltage of 200 kV. (See Note 2 for details.)

processing or growing semiconductors (such as silicon); large
ingots called boules, nearly a yard long and 6 inches in diameter,
are sliced into thousands of wafers onto which integrated circuits
are developed. Ingots of metals and alloys, or even castings formed
by pouring molten metals and alloys into special molds, solidify
by forming a large number of individual crystals in the melt, which
grow in size and grow together. This is a process almost identical
to forming rock candy on a string by allowing crystals to form
around the string from a saturated sugar solution. In the case of
metal or alloy ingots, however, these ingots can be further shaped
or forged or rolled into bars or plates for industrial and commercial applications.

    Figure 5 shows this arrangement of crystals (called grains) in
a commercial metal that has been ground flat, polished, and
chemically etched to exaggerate this arrangement. This analytic
process, called optical metallography,[1] is a commonly employed
method for examining the grain structure of a wide variety of
metals and alloys.

CRYSTAL IMPERFECTIONS, DEFECTS IN MATERIALS, AND PHASE EQUILIBRIA

If we reflect on the solidification process, it should be readily
apparent that in addition to the irregular growth of many crystals
from the melt, impurities and other irregularities, such as gases,
can become entrapped within the solidifying masses. Indeed, many
of these features can occur, and it is important to control the
solidification process in order to prevent undesirable occurrences
and to facilitate desirable ones. That is, it is not always desirable
to have a pure material, a single crystal, or a single, homogeneous
phase. In many instances, the properties and behavior of a material
can be significantly and advantageously modified by the addition
of a submicrometer ceramic powder to a metal during solidification or consolidation to form a composite, providing increased
strength over the metal itself, or through the precipitation of a
second phase within the material to provide the same kind of added

strengthening. Figure 6 shows thorium dioxide (ThO_2) particles having a distribution of sizes dispersed in pure nickel. These particles are actually single crystals of ThO_2 and the nickel is said to be dispersion hardened as a result of this particle dispersion. In Figure 6b, NiAl precipitates (see Figure 4) have been formed in an iron-chromium-nickel-aluminum alloy (Fe + 12 percent Cr, 5 percent Ni, and 5 percent Al). These precipitates are also single crystals and can strengthen or harden the alloy matrix, properties demonstrated by the dispersed particles in Figure 6a. The particles provide additional strength by making it more difficult for the matrix to deform. (We describe this deformation process in more detail later.)

Whether a second phase in a material, as shown in Figure 6, is a desirable or undesirable feature is a subtlety that may depend upon the application of the material. For example, the grain boundaries or regions separating one crystal from another in Figure 5 are defects because they interrupt the regular crystal lattice arrangement of either crystal separated by this interface. It turns out that these interruptions, however, like the particles in Figure 6, can hinder the deformation of a material. In fact, in many metals and alloys, the finer the grain size (and conversely the greater the number of such grain boundaries), the greater is the improvement in strength.[1] On the other hand, the grain boundary regions (regions separating one crystal from another) can be weaker than the rest of the material (each individual crystal), and the material can be pulled apart by separating at these boundaries just as a foam coffee cup can be pulled apart by breaking along the individual foam grains. This process, called intergranular brittle fracture, is exacerbated by impurities or foreign atoms that, during solidification or heating, migrate to these interfacial regions and allow the material to break apart easily. This process is known as grain boundary segregation.[3]

It is apparent that if defects, precipitates, segregation of impurities, and the like can be controlled in materials, their behavior can be controlled to some extent, or at least their applications can be controlled or limited. This is a crucial feature of engineering design.

The processing of iron and steel is perhaps one of the best examples of the ability to control a complex array of materials properties. This is one reason for the popularity of iron and steel in a host of industrial, commercial, and engineering applications. For example, iron containing different concentrations of carbon (up to 5 percent) characterizes a range of steels starting with the cast irons with high carbon content and moving to high-strength tool steels with medium carbon contents and low-carbon specialty steels. In many specialty steels other elements are added to the iron for a variety of applications, including of course the wide range of stainless steels. The carbon allows a whole range of microstructures or other phases, including variations in crystalline structure (bcc and fcc). This is determined by phase equilibrium considerations and can be understood in large part by studying phase equilibrium composition diagrams.[1,4] The phase equilibrium composition diagram for the iron-carbon system is shown in Figure 7. On heating, iron undergoes two so-called polymorphic phase transformations. Below 910°C it crystallizes as a bcc (a) structure. Between 910°C and 1390°C, it crystallizes as an fcc (γ) structure. (Refer to Figure 4.) Between 1390°C and the melting point, iron crystallizes in another bcc structure referred to as the δ-phase. (The a-Fe is called the ferrite phase; the γ-Fe is called the austenite phase.) The magnetic properties of iron are interesting. However, only the ferrite or bcc phase is magnetic. In fact, the ferrite phase goes through a magnetic transformation at 768°C; above this temperature iron is generally not magnetic.[5]

Referring again to Figure 7, observe that at 1300°C, the iron may absorb about 5 percent carbon. As the liquid cools to 113°C, cementite (iron carbide, Fe_3C, consisting of 93.33 percent iron and 6.67 percent carbon) crystallizes and separates. Cementite is very hard and brittle. It is in fact the hardest constituent of plain carbon steel and has a tetragonal crystalline structure.[1] Cementite is metastable and decomposes into iron and graphite. In most slowly cooled cast irons, graphite is therefore an equilibrium constituent at room temperature. An iron-iron carbide eutectic (containing 4.2 percent carbon), called ledeburite, also forms, as shown in Figure 7. Obviously, the formation or mixing of these different

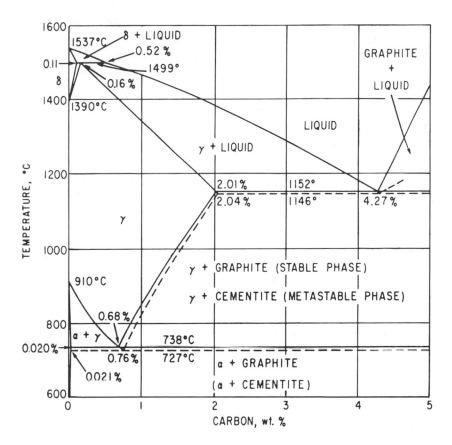

Figure 7. Iron-carbon phase equilibrium diagram. The dashed line represents the phase boundary for metastable equilibrium with cementite (Fe₃C). (From *The Making, Shaping, and Treating of Steels*, U.S. Steel Corp.)

phases has a profound influence on the properties and behavior of carbon steels, and these can be altered by proper cooling and heat treatment. These phases can also be controlled by the addition of other elements to the melt. For example, silicon additions can influence the decomposition of cementite to form graphite, and sulfur additions can retard graphitization in castings. A fine lamellar aggregate of iron and cementite, called pearlite, can also form in carbon steels. This microstructure resembles mother-of-pearl when viewed by optical metallography, and its name is derived from this appearance. Cast iron is therefore dramatically influenced in terms of its residual microstructure and properties by heat treatment and cooling rate. The most dramatic comparisons can be made between cementite, which is very hard and brittle, and graphite, which is very soft and flaky. Cast iron parts or commercial items can therefore possess dramatically different properties. Among the common cast iron uses are pipe, ingot molds, engine blocks, and the like. The following case example illustrates this feature.

Case Example 2

A worker in a refrigeration plant sustained a serious eye injury when a valve in a pressurized liquid ammonia system cracked open, spraying ammonia into his face. The valve was cast iron, and workers were instructed not to heat the valve. Heating of the valve, however, was a common practice in order to purge impurities that build up in the refrigeration lines.

During litigation of the injury claim, the prosecution speculated that the valve was somehow defective and this defect caused a catastrophic crack to form, as shown in Figure 8. A section of the valve crack was removed from the valve in order to examine the fracture surface in detail in the scanning electron microscope. This test piece was also polished and electroetched along the cut surface in order to examine the microstructure by optical metallography (refer to Figure 5). Figure 9 shows the typical results of these microscopic examinations. In Figure 9a, the fracture surface along a section of the crack exhibits very smooth separations containing flaky films. Figure 9b shows the microstructure to be characterized by very prominent and large graphite flakes in the cast iron matrix. There is therefore a strong correspondence and correlation between the fracture surface characteristics and the microstructure; the

Figure 8. Catastrophically cracked cast iron valve in ammonia refrigeration line.

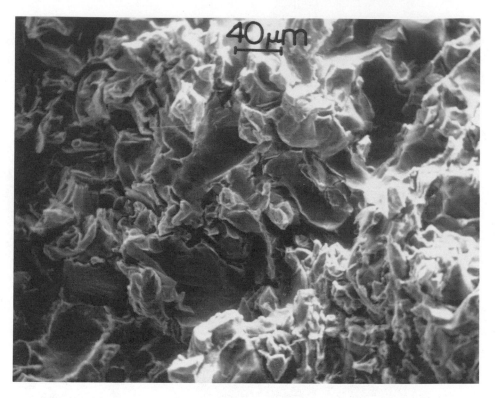

(a)

Figure 9. Microscopic examination of fracture surface (a) and
internal microstructure (b) in a crack section from the failed cast
iron valve shown in Figure 8. (a) Scanning electron microscope
view of a typical section of the fracture surface. Note the smooth,
flaky sections, which appear to be tearing apart and defining the
crack propagation path.

fracture apparently followed the graphite phases, which have
little or no tensile strength.

On the basis of the evidence summarized in Figure 9, it
was concluded that, in this application, slowly cooled cast
iron valves constituted an engineering design risk because the

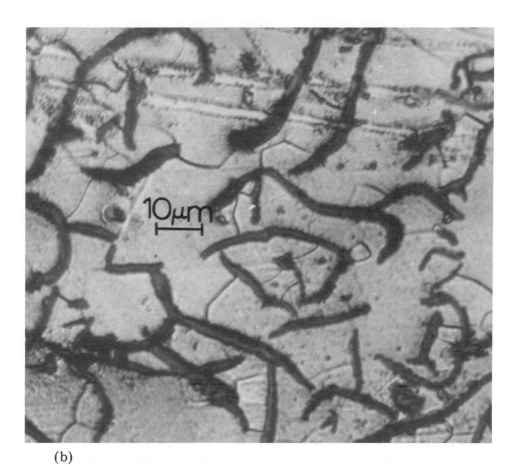

(b)

Figure 9. (Continued) (b) Optical metallograph view of microstructure in the cast iron valve. The thicker black regions are
graphite flakes, which constitute a fairly significant fraction in the
cast volume.

drastic variations in temperature the valve might routinely
experience (without unwarranted heating of the valve) would
be conducive to crack formation. Cracks would be expected
to form at the graphite/pearlite interface because of the
dramatic differences in their thermal volume expansion

coefficients (about 2 orders of magnitude smaller for graphite than for the iron-rich matrix). Once a crack is formed, its propagation would be expected to be very easy and catastrophic by running through the high-volume fraction of graphitized phase regions.

This may be an arguable conclusion, but as is typical of many evidenciary arguments, the claim was settled in favor of the plaintiff. The action in this particular case may also have been taken to prevent an examination of the entire engineering design criterion, which then might have been tested in a court action.

The iron-carbon phase diagram in Figure 7 is referred to as a binary phase diagram, and when three elements or components are involved, a more complex ternary diagram provides a "map" showing the ranges of compositions and temperatures within which various phases and microstructures are stable and the boundaries at which phase changes occur. More complex phase diagrams can also be constructed, but many combinations of two elements follow much simpler binary mappings.

CLASSIFICATION OF IMPERFECTIONS

Figures 5 and 9 illustrate various microstructures and phase regimes in materials that might be considered imperfections or volume defects, but they are controllable features of the microstructure. These features are certainly not the only imperfections, however. In the context of fundamental imperfections, there are four major classes: point imperfections, line imperfections, planar or two-dimensional imperfections, and volume or volumetric imperfections. Figure 10 illustrates these imperfections. Point imperfections, corresponding to single-atom irregularities, do not make too much sense within the definition of an amorphous material, as illustrated in Figure 2b. The same is true of line imperfections or dislocations, which require a regular crystal lattice as a basis for definition. It may also be questionable that

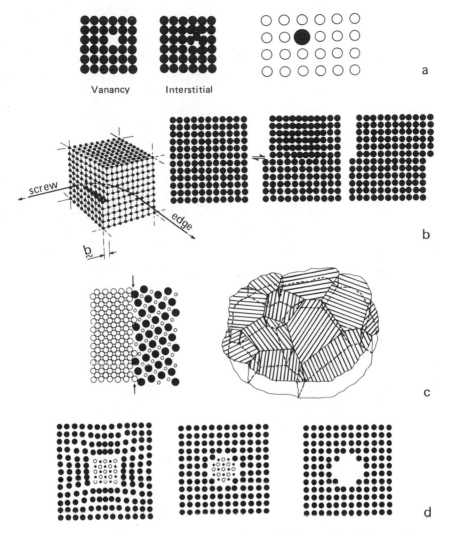

Figure 10. Fundamental classes of crystalline material imperfections: (a) point imperfections: vacancy, interstitial, substitutional impurity; (b) dislocation (line) imperfection and dislocation slip to create lattice displacement. The displacement created by a single dislocation is shown as *b*, called the Burgers vector. (c) Planar imperfections can be envisioned as interfaces separating crystalline grains or phases. (d) Volume imperfections: coherent precipitate (precipitate lattice is coherent with the matrix lattice), noncoherent precipitate or phase, void. Volume imperfections can of course be reconciled in amorphous solids as well.

31

planar imperfections, such as grain or phase boundaries, have any meaning in amorphous materials. Certainly crystalline regimes or phases occluded within an amorphous volume could be characterized by a crystalline/amorphous interface. To the extent that the interface characterizes the interruption of random arrangements or regular arrangements of atoms, this may indeed constitute an imperfection. Volume imperfections can of course be reconciled in both crystalline and amorphous materials. Collections of vacant lattice sites in crystalline materials or groups of substitutional impurities forming precipitates or other short-range phases are obviously easy to reconcile. Gas bubbles, voids, and clusters of impurities, however, can form in both crystalline and amorphous matrices. The effect of volume imperfections, particularly large volumes, can have a similar effect in both crystalline and amorphous materials. In any case, imperfections in materials can have a significant if not a controlling influence on the properties and behavior of many materials. Consequently, it is important to understand the nature and role of imperfections as a prerequisite for control based upon predictability.

Since we are particularly concerned in this presentation with deformation, especially the fracture of crystalline metals and alloys with practical commercial significance, it is necessary to develop the concept of dislocations because dislocations are in fact the principal mechanism of deformation. To understand this concept better, we might consider the consequences of deformation in common metals and alloys, such as aluminum, copper, aluminum-copper alloys, and stainless steel. All these materials possess the face-centered cubic crystal structure. A copper or aluminum wire can of course be easily bent or folded and, if of sufficiently small diameter, tied into knots. The same thin wire can be stretched considerably (by more than 30 percent) without breaking, and when so stretched, the wires will remain permanently stretched. Obviously this is not a matter of atomic bond stretching. What happens in fact is that atoms are made to be translated relative to one another so that planes of atoms in the most intimate or closely packed atomic plane arrangements are made to slip or glide relative to one another. Consequently, the process of creating an extra half-plane by systematic shear, as shown in the creation of a dislocation in Figure 10, which can glide through a crystal to produce an offsetting step, can be made to repeat with increasing frequency as the deformation (or stress level) is increased. Note that in Figure 10 the dislocation line connects the edge and screw components in the deformed volume element or is a line perpendicular to the page at the bottom of the extra half-plane of atoms.

This slip, as depicted atomically in Figure 10 and multiplied hundreds of thousands of times in a large structure or even a small wire pulled in tension, will allow a combined and permanent displacement in the direction of the applied stress in the same way a deck of cards can be sheared into a permanent displacement. Figure 11 illustrates this slip phenomenon for an idealized single-crystal rod pulled in tension. Corresponding to this combined slip, which can be regarded as the combined motion of multitudes of dislocations, as shown in Figure 10, a net displacement or extension of the rod occurs that corresponds to some applied force P or stress σ. This displacement is regarded as a net strain (as illustrated by ϵ in Figure 11), and when it is plotted against the corresponding load or stress (load divided by the cross-sectional area), a stress-strain diagram results. The stress-strain diagram describes the deformation history of a material in terms of the applied stress and is a principal feature of design strategy for most structural materials applications. The strain can be expressed as the percentage of elongation compared with the original length, or in dimensionless units as (total length after straining − original length)/ original length.

The concept of dislocations in illustrating the slip process provides a very precise description of the deformation process, especially the mechanical behavior of metals and alloys. Whenever a metal or alloy is deformed in any way by a stress that exceeds the yield stress illustrated in the stress-strain diagram of Figure 11, regions within the material respond as shown schematically in the dislocation-slip sequence of Figure 10. In practical, polycrystalline metals and alloys, the dislocations created in each grain may not be able to slip out of the crystalline grain or out of the materials as illustrated in Figure 10. In this way, dislocations and slip

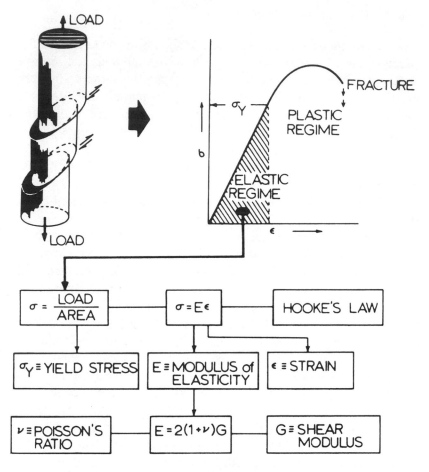

Figure 11. Deformation-induced slip in a single-crystal rod and
relationship to the stress-strain diagram for an idealized single-
crystal metal or alloy. The same effect and the same form of
stress-strain diagram is observed for practical, polycrystalline
materials. The elastic regime is a linear regime where Hooke's law
applies. A material stretched in this region will ideally relax to its
initial form when the load is released. The yield stress σ_Y represents a critical stress above which deformation will be permanent, and relaxing the load will not allow the material to completely recover to its initial form. Other simple mechanical
relationships are defined in the graphic components. (After Figure 2.22 in L. E. Murr, *Solid-State Electronics*, Marcel Dekker,
Inc., New York, 1978.)

displacements build up in a solid as deformation continues or as the stress is increased, and this creates a kind of back-stress that will make it increasingly difficult to deform a material. Sometimes called work hardening, this process can be appreciated by anyone who has bent a ski pole in a snowy spill and attempted to bend the pole straight, or the exhibitionist who on bending a large steel pipe with relative ease finds it virtually impossible to bend the pipe back to its original form. During this kind of deformation process dislocations created in the material become locked up or stacked against grain boundaries, making slip difficult. It should be apparent, since dislocations can interact with one another and with grain boundaries (planar imperfections), that they can be held up to some extent by any imperfections. So the obvious advantages of dispersed particles and precipitates in materials, as illustrated in Figure 6, are that they impede the movement of dislocations through a material and thereby prevent slip and deformation. The net result of course is that providing obstacles to dislocation motion can strengthen a material.[6]

Figure 12 provides a final example of dislocations, work hardening, and tensile deformation. In the sequence of bright-field transmission electron micrographs,[2] the buildup of dislocations with applied stress or strain is graphically apparent. It is of interest and of some value to note here that if the hardness of the strained stainless steel were measured corresponding to each of the strained images or corresponding to any increasing strain point, the hardness would be observed to increase with increasing strain. We will discuss hardness and hardness measurement in more detail in a later chapter. It is important to notice (in the context of Figure 12) that as the dislocation substructure increases in density and complexity with increasing strain, an indenter used to measure hardness by penetration and surface deformation would find increasing penetration resistance, which would be recorded in the hardness measurement as increasing hardness.

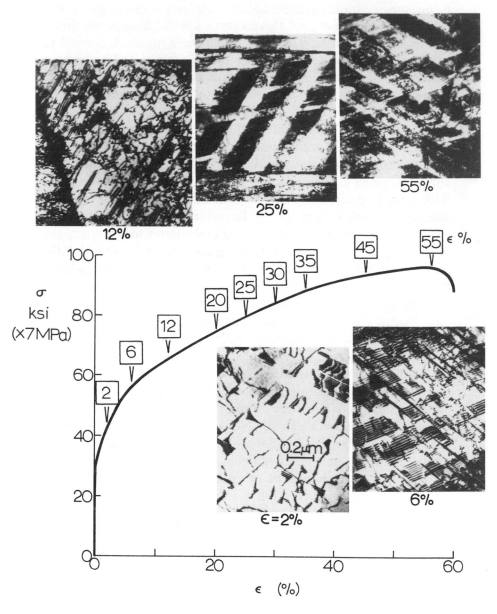

Figure 12.

Figure 12. Transmission electron microscope images of dislocation substructures in thin 304 stainless steel sheet samples corresponding to some of the strain levels in the stress-strain diagram. The black wiggly lines and fringes (especially at 2 and 6% strain) are imaged by systematic diffraction of electrons in the lattice. This phenomenon can be understood by referring to the total dislocation volume element in Figure 10. Note that the dislocations are slipping in specific crystalline planes and along specific directions in the fcc stainless steel grains. It is apparent that as the strain (and stress) is increased, the number of dislocations (crystalline line imperfections) increases dramatically. At high strains (above 12%), a transformation to a bcc (martensitic) phase occurs, and this contributed to the complexity of higher strain images. The stress axis σ shows values in thousands of pounds per square inch (ksi) or megapascals (MPa), which are given as equivalent units. [From L. E. Murr and S-H. Wang, Res Mechanica, 4, 237-274 (1982).]

3

Materials Characterization Using Light, Sound, X-rays, Electrons, and Ions

Light, sound, and X-rays can provide an overview of macrostructure but generally cannot detect microscopic or submicroscopic details. Electron and ion beams can provide a remarkable range of opportunities to characterize both the microstructure and the microchemistry of materials.

There are numerous probes and probe possibilities for viewing and examining materials. It would be impossible even to attempt to cover any or all of them in any detail. It is worthwhile to examine, even briefly, some of the more popular modes of characterization. We spend more time and develop more details for materials characterization using electrons and ions because the popular analytic techniques employing electron or ion probes allow imaging as well as rapid, high-resolution chemical or microstructural analysis.

LIGHT MICROSCOPY AND OPTICAL METALLOGRAPHY

It is likely that the reader is already at least phenomenologically if not functionally familiar with some form of light or optical microscope. There are two prominent modes of imaging matter using light: reflection microscopy and transmission microscopy. Obviously for very thick materials or opaque materials, light transmission has limitations. Certainly for metals and dense, optically opaque materials, only the surface can be imaged using reflected light. Image formation in a light microscope is accomplished by glass lenses shaped to focus the light rays.[7] The key to image formation is really the change in the velocity of the light waves as they cross the interface separating the air and the glass lens. The index of refraction is the standard measure of the relative velocities of light waves in any two media.

An optical microscope arrangement used for observations and studies of metals and other engineering materials that utilizes reflected light is generally called an optical metallograph. In this device, microstructure as viewed from the surface is observed by contrast developed as a result of differences in reflectivity from one phase to another. These differences are attributed to differences in surface texture or roughness resulting from chemical or electrochemical attack on different phases or at grain or phase boundaries, causing surface relief and texture changes that produce reflected light contrast, as illustrated earlier in Figure 5.

Optical microscopy or metallography provides a simple, direct overview of a material's microstructure as observed in a polished surface at magnifications limited to about 1000X direct. There are limitations to the depth of focus in a light microscope, and resolution is limited by the wavelength, which is nominally in the range of 0.2 - 0.6 micrometers, that is, the visible portion of the electromagnetic spectrum. Optical metallography can provide a very useful overview and a macroscopic perspective of a variety of failure analysis problems alluded to in part in previous chapters and discussed in detail elsewhere.[8]

It might be useful at this point to consider the role that wavelength plays in resolving features of microstructure in materials. It must be recognized at the outset that structure or microstructure cannot be resolved if it is smaller than the wavelength of the medium being used to form an image, regardless of whether the medium can be focused by a lens system. This is illustrated in Figure 13, which shows two pairs of spheres on a surface being imaged by reflected radiation. The wave nature and the parameter used in defining the wavelength of the radiation or imaging medium are illustrated in Figure 13.

It is certainly apparent from Figure 13 that light, particularly white light, does not allow molecules spaced a distance of only 0.002 - 0.006 micrometers to be viewed because their spacing is 100 times too small compared with the wavelength of light, as already noted.

ULTRASONIC INSPECTION AND ACOUSTIC MICROSCOPY

Acoustic or sound waves in liquids and solid materials can propagate without appreciable attenuation, and the generation and detection of acoustic waves at ultrasonic frequencies (and with corresponding wavelengths in the micrometer range) can be used to study underwater objects and internal structural or microstructural features in materials or organs inside the body. Ultrasound imaging of a fetus in the womb is a well-known example of this technology.

Although there are numerous techniques utilizing ultrasound for inspection, characterization, and imaging, the basic principle shown schematically in Figure 14 should suffice to illustrate a

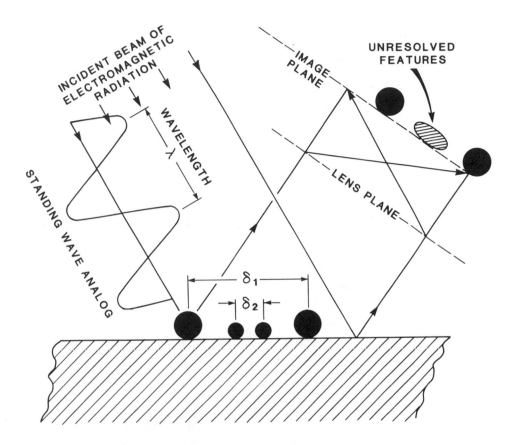

Figure 13. Conceptual relationship between wavelength and resolution in image features. Features in an object plane cannot be resolved in the image if the imaging wavelength is greater than the feature dimensions or displacement. Features are resolved where $\lambda < \delta_i$ ($i = 1, 2, \ldots$).

range of applications. In the technique illustrated in Figure 14, a transducer, consisting of a piezoelectric crystal or a similar sound wave generator, produces mechanical vibrations characterizing ultrasonic waves in a material. The transducer converts electrical impulses into mechanical vibrations, and vice versa, so that it both transmits and detects ultrasonic waves. By comparing the

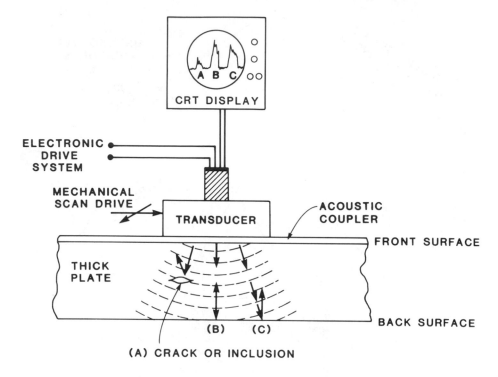

Figure 14.   Schematic illustration of acoustic wave inspection. Wave distortion caused by wave reflection from an inclusion or crack is differentiated from background signal pulses in a CRT scan. By scanning the acoustic wave synchronously with a CRT scan, such distortions can be imaged directly, as in ultrasound imaging as a medical diagnostic tool.

propagation and attenuation of ultrasonic waves in a material, structural changes are characterized by changes in the detected wave signals, as shown schematically in Figure 14.

As alluded to in Figure 13, if the acoustic wavelength is larger than the dimensions of the flaws to be detected, signal resolution disappears and the inclusion will go undetected. For large cracks, weld inclusions, and structural voids in steel plate, pipe, and a variety of materials, acoustic emission detection and ultrasonic inspection represent nondestructive testing methods for inspection and quality control.[9] Resolution is limited, however, and signal interpretation can be very difficult.

The acoustic waves generated as shown schematically in Figure 14 can be focused by fashioning a concave spherical interface below the transducer to form an acoustic lens.[10] This arrangement allows a focused acoustic wave or ultrasound beam focused to a spot in a plane of an object to be mechanically moved across this plane point by point and line by line in a raster pattern. Signals reflected from the object or features in the plane of the object are stored in an electronic memory and finally made to modulate the intensity of an electron beam in the CRT (cathode-ray tube) illustrated in Figure 14. By scanning the electron beam across the CRT screen, a process similar to producing a television image, synchronously with the scanning of the acoustic beam across the plane of the object an image is formed on the CRT in place of the signal peak shown in Figure 14. By displacing the electron beam across the CRT relative to the position of the acoustic beam in a material, the image can be magnified in proportion to the displacement.

Since the acoustic microscope is also limited in resolution by its operating wavelength, the ability to generate acoustic waves with wavelengths in the visible light range (0.2 - 0.6 micrometers) allows images to be formed with a resolution as good as that of the optical microscope. But since these ultrasonic waves can be focused with a variety of opaque materials, even thick materials, acoustic microscopy can be used as a unique diagnostic tool for detecting flaws in laminated materials systems, such as multilayer circuit boards[10,11] or imaging organs in humans and animals.

The acoustic microscope normally requires the acoustic coupler (Figure 14) to the object to be a liquid, and for many systems this can pose particular problems. In simple systems the acoustic lens is a tiny, concave spherical interface between sapphire and water.

RADIOGRAPHY, X-RAY MICROSCOPY, AND X-RAY METALLOGRAPHY

Radiography, utilizing energetic γ-ray or X-ray sources, has been used for decades as a means for examining and certifying a variety of materials, including thick steel plates and large welds in pipelines and nuclear reactor and other pressure vessels. In its simplest form, illustrated in Figure 15, an X-ray beam, considered to be radiating from a point source, penetrates a solid material and exposes a film sealed in a lighttight packet. Relatively large voids

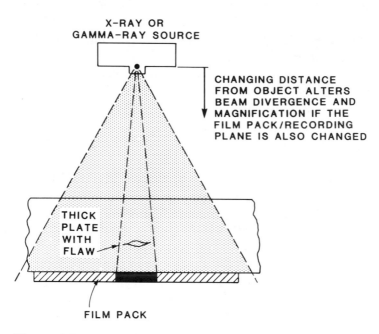

Figure 15. Schematic diagram illustrating the shadowgraphic X-ray detection of a flaw in a plate by projection of absorption differences in a photographic film. The shadow of a projected void will produce a darker region on the (negative) film in contrast to the background when developed. Magnification of the projection image is achieved by moving the film plane or the X-ray source, or both.

or flaws in the material provide contrast differences in the exposed emulsion because of differences in the absorption of the beam on encountering inclusions or voids, conversely allowing intensity differences to be used as a means for producing shadow images of the inclusions or voids. This is a very simple, nondestructive means of testing fabricated structures but lacks any resolution of small inclusions or defects even in the 100-micrometer range because there is no significant magnification factor.

X-ray wavelengths are less than the atomic spacing in metals, but because there is no convenient way to focus an X-ray beam, high-resolution and high-magnification images utilizing the technique shown in Figure 15 are not attainable. In fact, the simple X-ray optical system is not as efficient as a light optical system.

X-ray metallography, utilizing the fact that X-rays can be systematically diffracted from crystalline planes in a polycrystalline solid, has also allowed the structure, and even the defect microstructure, in metals and alloys to be examined.[12] A beam of X-rays reflecting (or more accurately, diffracting) from crystalline planes, as illustrated by the simple geometry that depicts Bragg's law in Figure 16, can allow the determination of the atomic dimensions of the unit cells composing a material (see Figure 4). Notice the similarity in geometry in comparing Figure 16 with Figure 13 and the role played by the wavelength. Since the X-ray wavelength would normally be known, X-ray diffraction from crystalline planes can allow the determination of the interplanar spacing depicted or the lattice parameters or unit cell dimensions. Figure 16 shows these calculations for cubic crystalline materials as depicted, for example, in Figure 4.

ELECTRON MICROSCOPY AND MICROANALYSIS

The techniques already described can provide rapid, nondestructive evaluation and characterization of a wide range of materials. They can provide an overview of macrostructure, but little or no direct information can be gained that can be related to chemical composition, with the exception of X-ray diffraction.

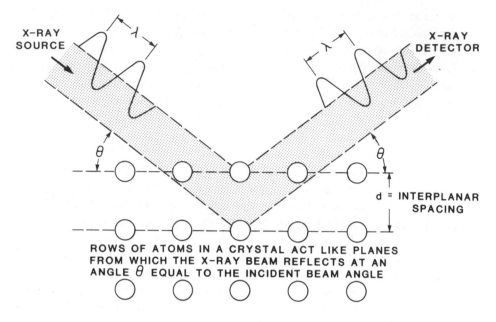

BRAGG'S LAW: $2d \sin \theta = \lambda$
FOR CUBIC CRYSTALS:

$$d = \frac{a}{\sqrt{h^2 + k^2 + l^2}}$$

a = LATTICE PARAMETER
OR UNIT CELL SIZE

h,k,l ARE INTEGERS DENOTING
SPECIFIC CRYSTAL PLANES

FOR THE CRYSTAL PLANES
SHOWN ABOVE h = 0, k = 1, l = 0
SO THAT d = a

Figure 16. Schematic illustration of the systematic reflection (or more accurately, diffraction) of an X-ray beam from crystalline lattice planes. The process and process geometry are described by a simple equation referred to as Bragg's law. The indices of crystalline planes are referred to as Miller indices, and the crystalline planes depicted here are the {010} planes; therefore, the interplanar spacing is the unit cell dimension; that is, d = a.

More importantly, these techniques lack any significant resolution and generally cannot image or detect microscopic or submicroscopic details.

An electron beam provides a remarkable range of opportunities to characterize both the microstructure and the microchemistry of materials. The use of electron beams must normally, however, be compromised against the necessity to destroy a portion of the sample, or at least to compromise its integrity in some way. Figure 17 illustrates the many reactions that occur when an electron beam impinges upon a solid materials. The resulting phenomena can be detected and treated as specialized, analytic signals.

As shown in Figure 17, a thin specimen in the path of an electron beam can be examined by detecting secondary electrons emitted from the material or primary electrons scattered (backscattered) from the surface of the material. Transmitted electrons can allow an examination of the internal structure of a material along with the diffracted electrons, which provide crystalline structural information in the same way that X-ray diffraction can provide information, as depicted schematically in Figure 16.

When an electron beam of sufficient energy interacts with a solid, X-rays are produced. In fact, if the source shown in Figure 15 were an X-ray source, it would consist of an electron gun that would generate a high-energy electron beam directed against a metal target to form X-rays. The generation of X-rays is a fundamental process involving the excitation of individual atoms. In this process, electrons composing the atoms are caused to jump from a lower energy state to a higher energy state. In Figure 17 the energy is provided by the primary electron beam. When these excited electrons return to original or lower energy levels, energy is released. This energy E is given by a simple formula

$$E = h\upsilon = \frac{hc}{\lambda}$$

which is called Planck's energy equation. In this equation, h is a constant called Planck's constant, υ is the frequency of X-radiation emitted by this atomic energy level transition, c is the speed of

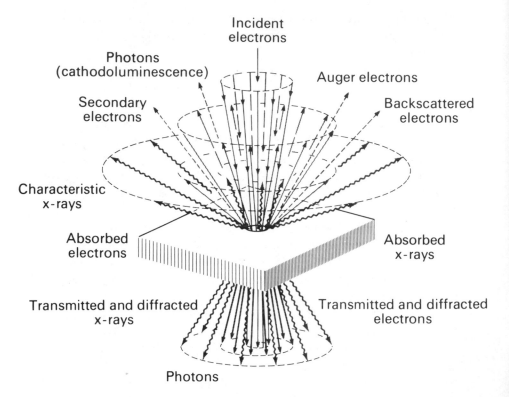

Figure 17. Schematic view of principal interactions and effects when a primary beam of electrons impinges upon a thin solid material. (After Murr, see Note 2.)

light, and λ is the corresponding wavelength of the X-radiation. Since the electronic structure is different for each atom, the X-ray emission will be different from each atom. The X-rays emitted from each atom are thus characteristic of that atom. Moseley's law states that the characteristic energy E of characteristic X-rays is proportional to the square of the atomic number $Z : E \propto Z^2$. If a sample contains a mixture of atoms and the characteristic X-rays can be detected and compared, the intensities of the X-ray signals can be used with high precision to determine the elemental compositions (the concentrations of Z atoms, where $Z = 1$ for hydrogen, $Z = 2$ for helium, $Z = 3$ for lithium, and so on, according to the periodic table of the elements.)

Since electrons are charged particles with wavelike qualities, they can be focused by forming a lens utilizing either electrical or magnetic fields. Magnetic fields have certain advantages for high-energy electrons. Not only can electrons be focused, but beams of focused electrons can also be deflected. This makes possible the operation of a television receiver in which focused electron beams are systematically and synchronously scanned and blanked across the face of the picture tube to form black-and-white or color images.

This same process, or raster scanning of the beam, is used to examine materials in a scanning electron microscope. By scanning a focused electron beam across an area of a specimen under examination, backscattered and secondary electrons can be detected and displayed synchronously on an imaging CRT to provide details of the surface topography and microstructure, as demonstrated by the secondary electron image of Figure 9a.

At the same time that an electron beam is scanned across an area of a material under examination, characteristic X-rays are produced at each point of location of the beam. Consequently, if an X-ray detector collects and differentiates this signal information, the elemental composition averaged over a single line scan or within a specifically scanned area can be displayed or compared. This mode of characteristic X-ray analysis is illustrated schematically in Figure 18.

The analysis of characteristic X-rays emitted from a specimen, as shown in Figure 18, is referred to as energy-dispersive X-ray (EDX) spectrometry. A spectrometer is any detector arrangement that can provide a spectrum of energies, wavelengths, or other parameters that define the identities and concentrations of elements (atoms) or molecules. The X-rays generated by scanning the electron beam over the sample or specimen surface in Figure 18 strike a silicon semiconductor detector containing lithium impurities to enhance the detection capabilities. Each X-ray interacting with the detector produces a voltage whose amplitude is in

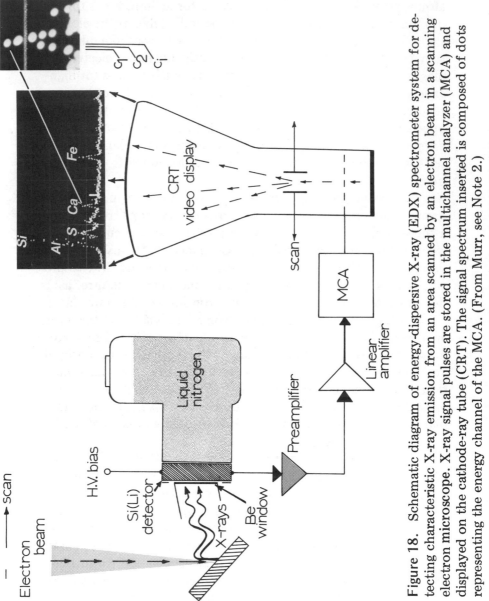

Figure 18. Schematic diagram of energy-dispersive X-ray (EDX) spectrometer system for detecting characteristic X-ray emission from an area scanned by an electron beam in a scanning electron microscope. X-ray signal pulses are stored in the multichannel analyzer (MCA) and displayed on the cathode-ray tube (CRT). The signal spectrum inserted is composed of dots representing the energy channel of the MCA. (From Murr, see Note 2.)

direct proportion to the X-ray energy. These voltages are amplified and stored serially in a computer with specific memory channels corresponding to a small but finite voltage range. When the detector is turned on for a specific period of time, the number of X-rays corresponding to specific elements composing the sample are therefore converted to a corresponding voltage, and as particular pulses are stored in memory and displayed, a spectrum results that depicts not only the elements present but also the relative intensities corresponding to relative concentrations. These can of course be accurately calibrated to determine the elements present with a precision of a fraction of 1 percent. Since the electron beam can be scanned over an extremely small area, an analysis can be obtained for an extremely small specimen. This technique would in fact have served the inquiry discussed in the example in the Introduction.

By selecting a narrow energy range within the memory channels of the multichannel analyzer in Figure 18 and using the signal detected to synchronously scan the electron beam of the video display CRT, it is possible to produce a map of particular element locations and relative concentrations at these locations. This technique can actually be used to selectively image elemental concentrations derived not only from characteristic X-rays generated in a sample, but also other characteristic signals, such as Auger electrons, noted in Figure 17, which, like X-rays, originate from specific atoms. Figure 19 provides an appropriate example of elemental maps over a small area of the fracture surface of a cast iron. The example shown in Figure 19 attests to the ability to very selectively image and analyze very small inclusions and impurities in a fracture surface.

In addition to electron microscopes and microanalysis systems that scan an electron beam across a small area on a specimen surface (Figure 20a) as a means to provide image signals and characteristic emission signals for micro-analysis (Figure 17), transmission electron microscopes also allow direct imaging of the internal microstructure in very thin films of materials using either a stationary or a scanning primary electron beam. Figure 20b shows for comparison the general appearance of a transmission electron microscope.

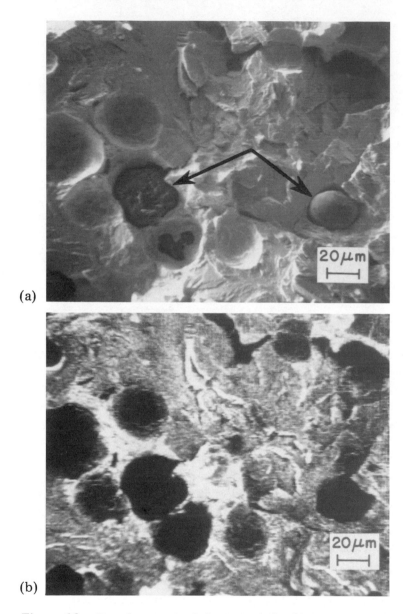

(a)

(b)

Figure 19. Development of characteristic element maps at precipitates and other features in a fracture surface of a cast iron by detecting Auger electron signals. (a) Secondary electron image of a surface region showing precipitates and other fracture surface features. (b) Iron map showing regions (dark) that are not iron or iron rich.

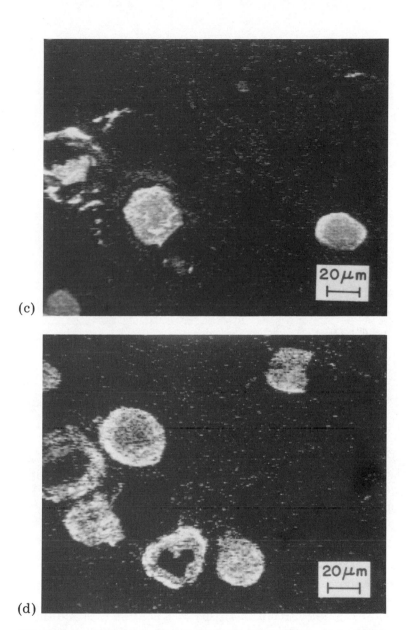

(c)

(d)

Figure 19. (Continued) (c) Carbon map showing precipitates illustrated by arrows in a to be carbides or carbon-containing compounds. (d) Antimony map showing concentration or segregation
of antimony to specific regions of the material. (Courtesy of Dr.
A. Joshi, after Murr; see Note 2.)

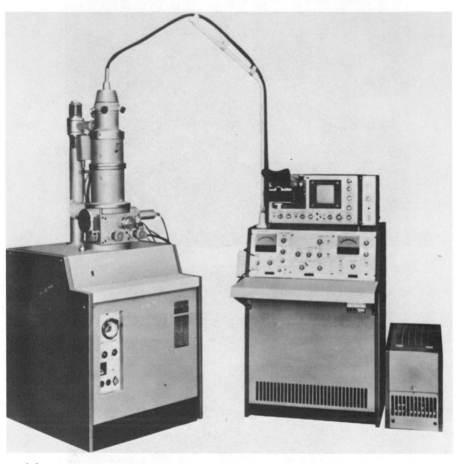

(a)

Figure 20. Typical views of electron microscopes. (a) A commercial scanning electron microscope showing the electron optical
column on the left (which is nothing more than a large, metal
black-and-white television tube sitting on a console top) and the
image display and electric control system on the right. (b) A commercial transmission electron microscope on the right is shown in
schematic cross-sectional view. Note the similarity of the electron
optical columns for both machine views. (Courtesy of JEOLCO
USA, Inc.)

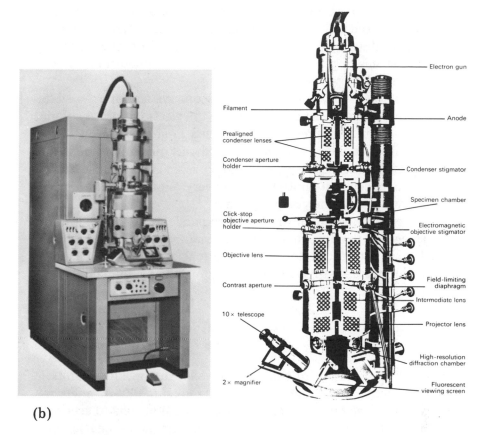

(b)

Figure 20. (Continued).

A comparison of Figure 20a and b certainly illustrates an obvious structural similarity. Indeed, either instrument might be considered to appear like a large metal black-and-white television tube sitting screen-down on a table surface. The top of the tube is connected to a high-voltage cable, which accelerates electrons generated in the electron gun into a series of focusing lenses. In the scanning electron microscope arrangement of Figure 20a, the beam is focused and scanned across the surface of a specimen. In Figure 20b, a focused electron beam penetrates a very thin specimen and is refocused in an objective lens to form an image of the internal microstructure. These transmission electron microscope images in thin metal or alloy specimens appear as shown typically in Figures 6 and 12. Figure 12 actually shows the images of dislocations (illustrated schematically in Figure 10) that represent rows of atoms displaced by an applied stress in deforming the stainless steel alloy.

In state-of-the-art electron microscope systems, it is possible to attach a range of signal detectors to the electron optical column to allow microstructures to be examined in great detail in thin films while obtaining very detailed and very precise chemical information. Such systems, referred to as an analytical electron microscope, can cost $500,000 at the very minimum but provide an unprecedented analytic precision.[2]

A critical feature of the precision possible with an electron microscope is the dependence of the primary beam wavelength on the voltage used to accelerate the electrons created in the electron gun. This wavelength is given approximately by the simple formula[2]

$$\lambda \cong \frac{12.3}{\sqrt{V_o}}$$

where if V_o is in units of volts, the wavelength λ will be in units of Angstroms; where 1 angstrom unit is equal to 0.0001 micrometer. Many large television sets operate at 30,000 volts to accelerate electrons from the gun to form a beam that is scanned across the screen to form an image. Most commercial transmission electron microscopes by comparison operate at 100,000 volts. Substituting this value for V_o in the electron wavelength equation will result in a wavelength value λ of 0.037 angstrom. Remembering that the average spacing of atoms in metals and alloys is around 3 angstrom, and referring to Figure 13, the reader should be convinced that ideally the resolution potential of a transmission electron microscope is about 100 times better than actually needed (compare λ = 0.037 angstrom with d = 3 angstroms in resolving atoms in a metal or alloy in the context of Figure 13). Obviously, working at higher voltage allows an even greater theoretical or idealized resolution, and when viewing thin metal or alloy films, the high voltage will provide greater penetrating power for the electron beam.

It is in fact possible actually to see atoms in an electron microscope, but this is not as easy as one might expect. The reason is that the magnetic lenses used to focus the electron beam in the image-forming objective lens (Figure 20b) are not perfect, just as many camera or other glass lenses for optical beams are not perfect. So lens and other machine distortions reduce the ideal resolution to the point at which it is just possible in modern instruments actually to see atomic or molecular detail. This will of course improve with improvements in the instrument.

A final advantage of electron microscopy over many other microscopic imaging systems, particularly light, is the large depth of field or depth of focus. In a light microscope operated at very high magnifications (only in the range of 1000X), certain features of a fracture surface will be in focus but others will be out of focus. In addition, light will scatter from very fine surface microstructures and contribute to image blurring. In the scanning electron microscope (SEM), these surface features are not blurred by scattered electrons and the microstructure will be in focus at 1000X, 10,000X, or even 100,000X. This feature provides a depth perception exactly akin to our visual capabilities, but applicable to the most microscopic and even submicroscopic features of an object.

ION MICROANALYSIS

Figure 21 shows, in comparison with Figure 17, the interaction of an ion beam with a solid material. The ion beam interaction phenomena are similar to electron beam interaction phenomena, but there are fundamental differences. One of the most significant features of ion beam interaction is the destruction of the material through ion beam sputtering or the systematic removal of the material in the interaction zone by the creation of secondary ions. In addition, even the simplest ion (hydrogen ion as H^+, or a proton formed by removing the single electron from a hydrogen

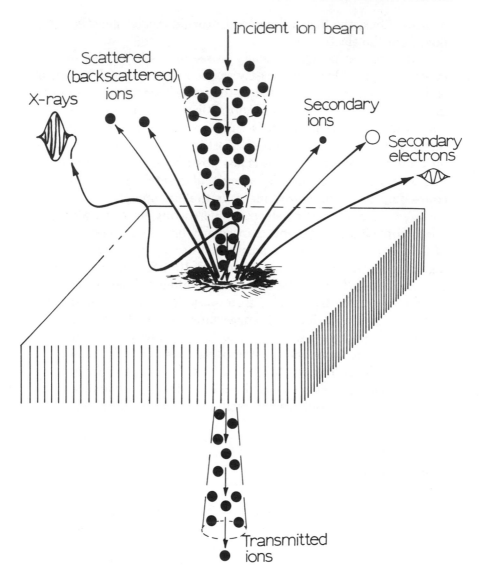

Figure 21.   Schematic view of principal reaction and interaction
features of a primary ion beam encountering a thin solid material.
(From Murr; see Note 2.)

atom) is nearly 1900 times heavier than an electron. This enormous mass causes extensive damage in transmitting a proton beam through a thin specimen of material. Ideally, however, since the mass of a proton is so much larger than that of an electron, the corresponding proton wavelength will be smaller than that of an electron by the mass ratio (proton mass/electron mass = 1837).

Ion beams, as charged particle beams, can be focused and deflected just as can an electron beam. However, because of the enormously heavier mass of ions relative to that of electrons, electrostatic lenses accomplish much more efficient focusing of ion beams accelerated with energies or voltages in the range of those used to accelerate electrons in an electron microscope or a related electron optical system. Furthermore, when mixtures of ions are accelerated through a magnetic field at right angles to the beam direction, the ions in the beam will be made to describe a radial path through the magnetic field that will depend upon the specific ion mass. Consequently, a magnetic field can be used to create an ion spectrum or selectively eliminate specific ion masses to act as a mass filter. This process is the basis for mass spectrometry or spectroscopy and is illustrated schematically in Figure 22. The ion source illustrated in Figure 22 could in principle be a mixed beam generated from a sample material impacted by a primary ion beam. By using a fixed aperture, ions can be selectively directed through the aperture by changing the magnetic field intensity to direct different mass values along a constant radius of curvature.

By combining some of the features of a scanning beam system with selective ion detection, a scanning ion microprobe can be devised that will allow very precise determination of a material's composition by the precision analysis of the secondary ion (specimen) mass spectrum. Figure 23 illustrates this analytic arrangement. This arrangement can also serve as a scanning ion microscope in which images can be formed by using either specific secondary ions or secondary electrons created by the primary ion beam interaction with the specimen materials. When images of the scanned surface are created using specific ions filtered from the secondary ion signal, an elemental concentration map results that

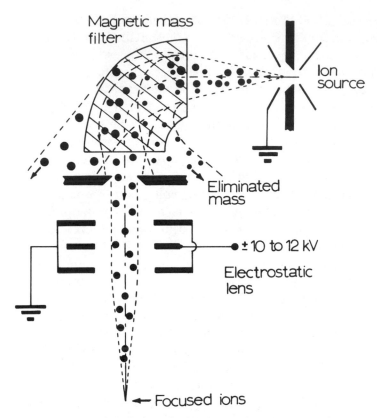

Figure 22. Schematic view of mass separation or mass filtering by a curved magnetic field region. By selectively changing the magnetic field, the ion selection can be changed and focused onto the detector system. (From Murr; see Note 2.)

is similar to an elemental X-ray or Auger electron map like those shown in Figure 19.

A simple but high-resolution microscope, the field-ion microscope,[2] can be used to image the surface atoms of a very tiny whisker specimen, as shown, for example, in Figure 2c and d. In this instrument a very high electrical field applied between the whisker specimen and a fluorescent screen causes gas atoms entering the system to be ionized at atoms over the whisker surface.

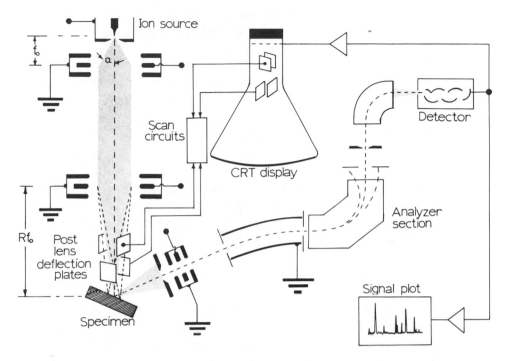

Figure 23. Schematic representation of a scanning ion micro-probe or mass spectrometer. The ion selectivity can be used to create characteristic element maps similar to those in Figure 19 or to produce a full spectrum of the elemental or molecular composition similar to that shown in Figure 18. (From Murr; see Note 2.)

These ions will follow electrical field lines extending from the surface atoms to the screen and cause a corresponding illuminated spot on the screen. The extension of these electrical field lines from the whisker surface to the screen provides enormous magnification, and by cooling the whisker to temperatures of several hundred degrees below zero (Fahrenheit), the movement of the surface atoms by solid thermal vibration is reduced so that the positions of individual atoms are resolved in the image, as shown in Figure 2c. This is a very specialized microscopic technique, applicable only to a special group of very hard metals and alloys that can be prepared in whisker form.[2]

Let us examine, retrospectively, the analytic example discussed in the Introduction in light of the electron and ion microanalysis techniques described here. Recall that a very tiny metal fragment had injured the eye of a worker and the fragment had been recovered but was destroyed in an attempt to perform a wet chemical analysis. This same fragment could obviously have been examined in a scanning electron microscope fitted with a nondespersive X-ray spectrometer system, as shown schematically in Figure 18, to obtain a characteristic X-ray spectrum of elemental composition, or an Auger electron detector could have provided similar spectral data that could have been stored in memory for comparison with a suspected source material or reproduced for similar comparisons. The EDX system illustrated in Figure 18 is limited to detecting elements above oxygen in the periodic arrangement of the elements ($Z > 8$) because of the silicon detector noise. The detector noise would even raise this elemental sensitivity were it not for the liquid nitrogen cooling, which suppresses the detector noise. Auger electron detector systems, although more cumbersome, do allow elemental detection down to hydrogen ($Z = 1$). However, the elemental resolution is only in the range of about 0.5% by weight on a routine basis. Nonetheless, these techniques, in the context of the small eye fragment, are non-destructive. The fragment could be re-examined by other laboratories or experts.

The same eye fragment might also be examined by some form of ion mass spectrometry, as illustrated conceptually in Figures 22 and 23. In order to analyze its composition by ion mass separation, however, a portion of the specimen fragment would have to be sacrificed to provide some quantity of secondary ions representative of the total sample composition. The resolution of this analytic process can be extremely high, easily orders of magnitude greater than the EDX or Auger spectrometer systems. Under certain circumstances only a small volume of such a sample specimen need be lost in the ion mass analytic process.

It must be reiterated that the options available in accurately characterizing materials, especially materials systems and components that have failed and that constitute the focus of litigation of liabilities, must be very carefully considered. The overview presented here is not exhaustive and was intended to develop some sensitivities to, and appreciation of, the level of analytic sophistication even routinely available to investigate a material's microstructure and microchemistry and their relationship to its properties and ultimate or actual behavior.

4

Materials Testing, Failure Analysis, and Fractography

Materials testing or mechanical testing involves methods for systematically measuring and evaluating the mechanical properties of materials. An examination of fractures utilizing engineering practice and characterization tools can lead to a reconstruction of a material's use based upon evidence obtained through an analysis of failed remnants.

We now come to the final areas of consideration in examining metal and material failure and its connection with establishing certain contingent liabilities. We have already dealt with some elements of materials testing, failure analysis, and fractography — the systematic examination of fracture surfaces. For example, Figure 11 illustrates the engineering evidence obtained in a simple tensile test of a material, and Figures 9 and 19 show examples of failure analysis and fractography. Specifically, Figures 9a and 19a are examples of scanning electron microscopy (SEM) fractography in failure analysis.

We will expand upon these concepts a bit more in this chapter and then present a series of additional case examples to further illustrate methodologies of failure analysis.

Those of us with a long-standing "sweet tooth" who can remember back two or three decades can recall a white taffy bar about 5 inches long by 2 inches wide and about 1/4 inch thick. These taffy bars, when eaten indoors or in the summer, were sticky and could be pulled into much longer bars. In cooler weather, they could easily be snapped into pieces and more conveniently eaten. At certain temperatures, however, the bars could be slowly pulled or broken by a quick pull or snap. In looking at the surface of these broken pieces, they would appear different if they broke by very slow pulling at warm temperatures compared with the smooth breaks seen when broken at colder temperatures or by very quick snapping. Many metals and alloys behave this way. If they are stressed slowly they stretch, but if pulled quickly, they break. Similarly, they stretch better at elevated temperatures, becoming brittle at lower temperatures. Some materials can be bent or stretched by a slow stressing or deformation process but will shatter when struck by an object. In addition, many of us have experienced the ability to break a piece of picture frame wire by folding or pulling it, but folding it back and forth repeatedly a number of times causes it to break. These are all examples of materials behavior, particularly stress behavior, and represent elements of mechanical tests that can be systematically performed to evaluate how a material will behave. This behavior, the kind of stress, and the method of stress can often be accurately deduced by examining the surfaces of fractured pieces of material.

MATERIALS (MECHANICAL) TESTING

Materials testing involves methods for systematically measuring and evaluating the mechanical properties of materials. These properties broadly include a material's strength, ductility, stiffness or toughness, and hardness. Strength properties are related to the ability of a material to resist applied forces; ductility measures permanent changes of shape without rupturing. Stiffness and toughness are related to the ability of a material to store energy of deformation; hardness is best defined as the resistance to deformation. Tension tests provide basic strength measurements under uniaxial loading. Fatigue tests provide measurements relating to strength under cyclic (reversed) loading. Toughness is measured by an impact test involving a weight striking a specially prepared sample of a material. The minimum force or absorbed energy required to fracture or break the specimen is recorded. This will change with rates of applying the force and the temperature of the material, and both are either standardized or measured independently. Finally, hardness is measured by allowing an indenter with a variety of geometries to be pushed into the surface of a material by a known or standard force and determining the depth of penetration of the indenter. This depth can be related to the size of the indentation as observed on the surface when simple indenter geometries are utilized. In many instances, there is a correlation between strength properties and hardness, and the simplest test can be used to obtain some insight into a number of interrelated mechanical properties.

Figure 24 is a graphic depiction of the principal testing methodologies briefly described above. The extended figure legend describes the salient features of the methodologies and measurements. An interesting case history might illustrate the significance of these mechanical properties in failure diagnosis.

Case Example 3

A hang glider manufacturer was being sued for several similar accidents involving cable failure. In stress simulation calculations of the cable stresses, it was demonstrated that the flight stress probably never exceeded the yield stress of the cable alloy. The only reasonable explanation for the cable failure was the possibility of fatigue. A close examination of the cable fastener positions and telephotography of the area during actual flights revealed large resonant wind-induced vibration in the suspect cable similar to the automobile radio antenna vibrations we observe at various highway speeds and weather conditions. Indeed, adding a small stabilizer structure to prevent the cable vibrations eliminated further incidences of in-flight failure. The fatigue failure diagnosis was therefore confirmed.

This example can be readily appreciated by comparing the stress-strain diagram and the S-N diagram in Figure 24. The apparent feature to note is that the stress to fracture is dramatically reduced as the number of cycles increases. Since the hang glider cable flexure described in the example represented a high-cycle fatigue situation, a low-fracture stress led to premature and unwanted failure.

FAILURE ANALYSIS AND FRACTOGRAPHY

We have alluded to the fact that brittle and ductile tensile fractures are distinctly different and depend upon the rate of loading or stress, the magnitude of the stress, the regularity or irregularity of the stress (such as cyclic loading), and the deformation temperature. Figure 24 is phenomenologically indicative of these differences. Fundamental considerations are also different. For example, since as illustrated in Figures 11 and 12 dislocation motion and production are intrinsic features of ductile metal deformation, extreme brittle behavior occurs in the absence of any significant dislocation (slip) motion. This difference in internal microstructure would be expected to produce fundamental differences in the appearance of the fracture surface structure. The science of studying fracture surfaces is in fact the essence of fractography.

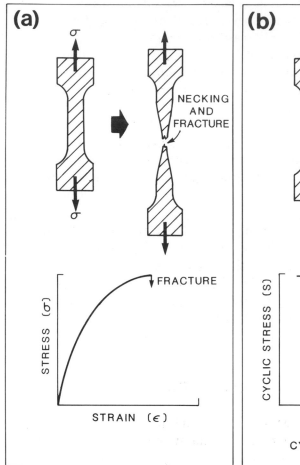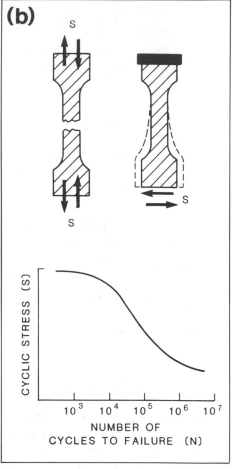

Figure 24. Schematic representation of principal mechanical testing methodologies. (a) Uniaxial tensile testing. Specimen "necks" as it stretches and finally fails. The corresponding stress-strain (σ-ϵ) diagram records this mechanical history. The fracture stress is often lower than the ultimate tensile stress, as shown in Figures 11 and 12. The details of this process are indicated in Figure 11. (b) Cyclic fatigue testing. The specimen can be pushed and pulled by equivalent cyclic stress S or clamped at one end and cycled by bending using a uniform stress S. The specimen is cycled at a fixed stress and the number of cycles at that stress recorded

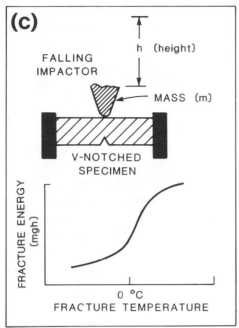

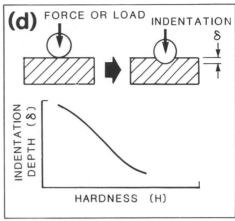

Figure 24. (continued)   when fracture occurs. This information is plotted in an S-N diagram shown. (c) Impact testing. The impact energy (impactor mass X gravitational constant X height the impactor falls) at which failure of a notched specimen fractures is recorded at different temperatures. Many materials are brittle at low temperatures and therefore fracture at very low impact energies as shown. (d) Hardness testing. Relative hardness is measured by comparing the indentation geometry or penetration depth. These measurements will also change with temperature. Hardness generally decreases as temperature increases.

It should be apparent that the appearance of structures and microstructures characteristic of a fracture surface has some relationship to the internal microstructure of a material and to the deformation history culminating in failure. An examination of fractures utilizing the engineering practices and characterization tools already described can provide some evidence, and even absolute proof, for the manner in which a material or a materials system was used. This is perhaps the essence of failure analysis — a reconstruction of a material's use based upon evidence obtained through an analysis of failed remnants.

Figure 25 illustrates schematically some of the fracture topographies to be associated with several particular fracture modes. Ductile metals and alloys begin to form voids at failure, and as these voids coalesce and pull apart a series of microscopic cups and dimples appear on the new fracture surfaces created (Figure 25a). In shearing a ductile material, these voids are squashed into elongated ellipsoids that form oval dimples or topographic tongues that lie in the direction of shearing (Figure 25b). These two fractography features are clearly illustrated in the examples shown in the scanning electron microscope images of Figures 26 and 27. Brittle metals or materials with reduced ductility will not form extended voids, as shown in Figure 25c, and the fracture surfaces are smooth separations, with some serrations formed by cleavage steps and similar phenomena. This feature is illustrated in Figure 28, which should be compared with Figures 26 and 27. The cleavage step edges and terraces over the surface are observed to run together in an interconnected series resembling tributaries and rivers; this feature is commonly referred to as a river pattern or feathermark fracture.

Figure 29 illustrates two other very common and distinctive fracture modes. Figure 29a shows a variation of cleavage fracture or brittle fracture that occurs between individual grains or crystals in certain polycrystalline materials. This brittle intergranular fracture looks like a foam coffee cup that has been pulled apart and often occurs when impurities or foreign atoms coat the boundaries or interfaces separating individual crystalline grains, causing weakening of the interatomic bonds or bond cohesion. The example

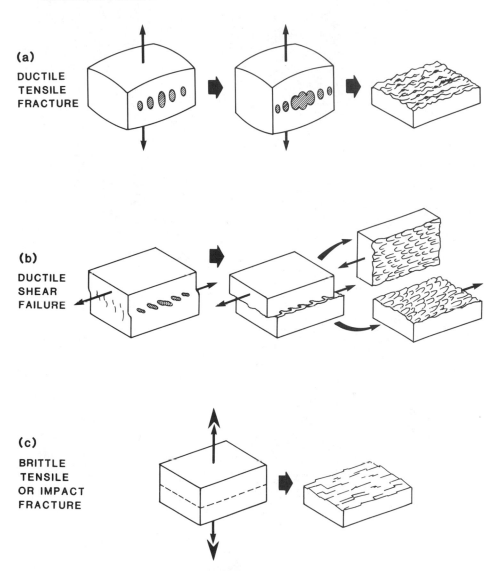

Figure 25. Schematic representation of the development of common fracture features. (a) Development of ductile tensile fracture features. (b) Development of ductile shear fracture features. (c) Development of brittle cleavage fracture features.

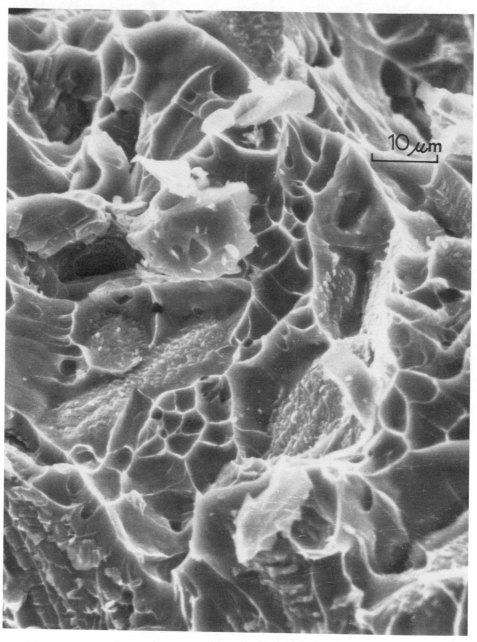

Figure 26. SEM fractography showing typical ductile tensile fracture surface features in a stainless steel.

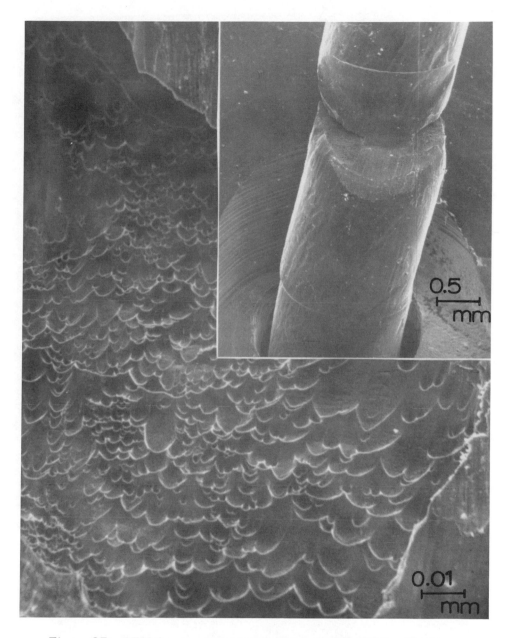

Figure 27. SEM fractograph showing an example of ductile shear fracture surface features in a titanium-molybdenum alloy wire. The wire is shown at low magnification in the insert. (Courtesy of Dr. N. M. Hodgkin, from Murr; see Note 2.)

Figure 28. SEM fractograph of brittle-cleavage fracture features
of a niobium-molybdenum-vanadium-zirconium alloy wire. A
smooth cleavage fracture area is shown perpendicular to the arrow,
which illustrates the wire axis direction.

shown in Figure 29a is rare for a face-centered cubic metal; this is a unique property of iridium at low temperatures (room temperature and several hundred degrees above room temperature).

Figure 29b illustrates the distinctive fracture surface topography and microtopography for fatigue failure. Because of the cyclic stress involved, movements and interactions of slip dislocations produce dense, somewhat ordered arrays of such microstructures within a material. When fracture begins these arrays produce surface ridges and microridges or striations as a crack grows. Investigations of particular metals and alloys have indicated that the striations are related to the stress cycles and the stress amplitude. Low-cycle fatigue produces larger striations; high-cycle fatigure produces very finely spaced striations that can sometimes only be seen at very high magnifications. Because of the regularity of the striations, such evidence for fatigue failure is nearly impossible to observe by optical (light) microscopy, not only because of the very fine detail, but also because systematic (spectral) reflection from the surface blurs the image detail. In fact, this is generally true of all microscopic surface detail, including ductile and shear fractography.[8,13]

In some cases of metal or alloy part failure, the fractographic evidence may involve several modes. For example, intergranular brittle failure may be visible, with fatigue striations observed on the fracture surfaces of some grains. One might then correctly assume that fatigue was involved in the deformation process, but if the material did not normally exhibit intergranular brittle fracture, additional considerations would enter into the failure analysis process, as demonstrated in the following case example.

Case Example 4

In a case involving the failure of a trailer coupler, a 10-ton grain combine was extensively damaged. Shortly before the incident, the combine owner had the coupler welded to his truck, and since the weld did not fail and the load was under the load rating of the coupler, which fractured at the base of the ball-pin assembly (the ball being 2 5/16 inches in diameter and the pin having a 1.19-inch diameter), the integrity of

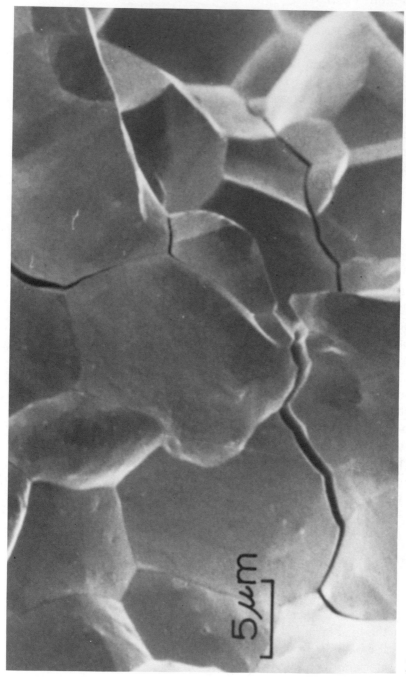

5 μm

(a)

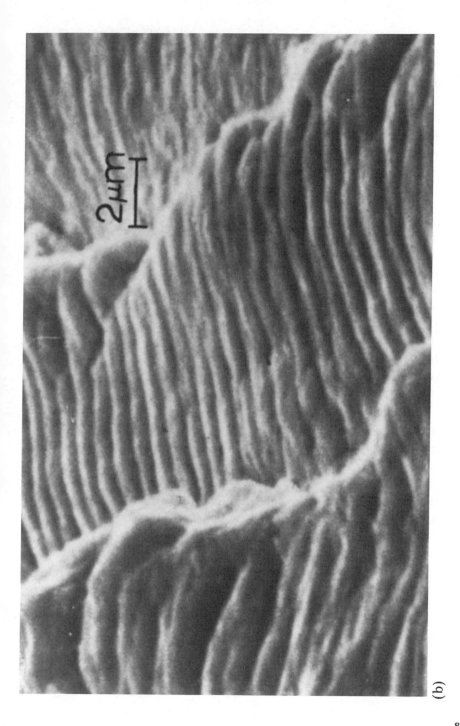

(b)

Figure 29. SEM fractographs of intergranular brittle fracture in iridium (a) and fatigue fracture in OFHC (oxygen-free, high-conductivity) copper showing regular fatigue striations as principal surface features (b).

the coupler was suspect and a suit filed against the distributor and the manufacturer. Engineering data for this drop-forged ball showed it to have a minimum tensile strength of 100,000 psi (pounds per square inch; popularly noted as 100 ksi, where ksi represents thousand pounds per square inch), a yield stress of 60 ksi, and an iron alloy composition containing 0.9 percent manganese and a maximum of 0.6 percent silicon.

Stress calculations based on the load at the time of failure indicated the maximum stress to be 45 ksi, well below the yield stress.

Several small sections were sawed from the fractured pin and examined in the scanning electron microscope; and Figure 30 illustrates some typical examples of fracture surface images or SEM fractographs observed. The fracture was very characteristic of brittle, intergranular fracture, as shown for example in Figure 29a. This steel alloy does not normally fracture in this manner. It is normally much more ductile, and the fractography should exhibit the features implicit in Figure 26.

Furthermore, on energy-dispersive X-ray (EDX) analysis of the fracture surface, areas typified by Figure 30 especially showed more than 3 percent silicon and 1 percent or more calcium. Calcium seemed to be concentrated near intergranular fracture surfaces and was thought to be segregated there. From this evidence and the fracture surface images, it was concluded that the alloy steel processing somehow retained some flux impurities (calcium in particular) in the original mill heats. This can occur when the very end of a ladle melt is poured and some of the flux and slag residue are incorporated in the melt. Calcium can selectively segregate along grain boundaries and cause embrittlement.

Although the evidence here is certainly not unambiguous and irrefutable, the argument was sufficiently strong to encourage an out-of-court settlement by the manufacturer.

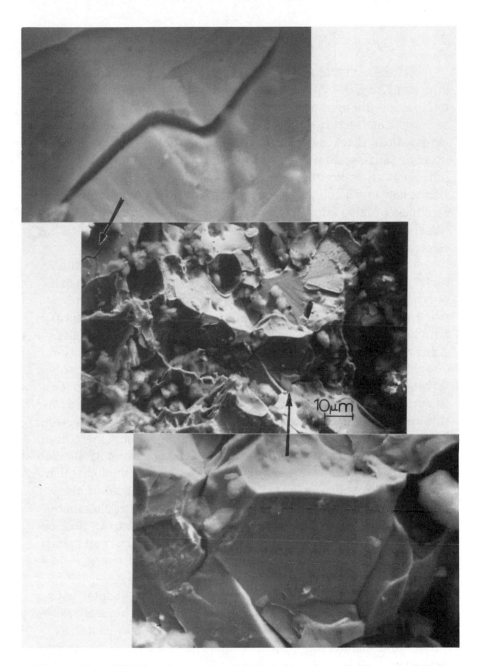

Figure 30.   SEM fractographs of failed drop-forged alloy steel ball coupler exhibiting uncharacteristic intergranular brittle fracture features.

Before concluding with some additional case examples, it is important to briefly discuss an important variation in electron microscope fractography. It is certainly apparent that direct observations of fracture surfaces in the SEM have many advantages. Nonetheless, there are circumstances that would not allow a fractured or failed system or system part or a fragment cut from a failed part to be examined in the SEM. A failed landing gear from an airplane crash and a failed drive shaft in a vehicular failure are examples of large samples that because of court requirements or litigation constraints could not be altered for analysis. Replication electron microscopy is a way around this.[2]

In replication microscopy, a "cast" is made of the fracture surface or a selected area on the surface by pouring a plastic or polymer setting liquid into the area or using a soft tape that when peeled from the area will contain all the intricate microstructural details.[2] Anyone who has peeled hardened airplane glue or the like from their fingers and observed the reproduction of epidermal microstructures can appreciate the process. The observation of these details can be enhanced by coating the replicated surface with a thin metal for observations in the SEM. In addition, very fine details can be observed in the transmission electron microscope (TEM) by coating the plastic replica with a metal film or a carbon and metal film (made by vaporizing carbon by striking a small arc between two graphite electrodes in a vacuum chamber and then following with a vaporized metal, such as gold or platinum). When this new film is made on the surface, it replicates the plastic or polymer replica or "cast." This original cast can then be dissolved in a suitable solvent, leaving only the thin carbon-metal replica film that can be observed in the TEM. The reader is referred to Note 2 for additional details.

We conclude this section and summarize this chapter with some additional case examples aimed at illustrating not only failure analysis and fractography, but also more generally materials science and engineering principles and materials characterization strategies.

Case Example 5

In a case involving an insurance claim settlement, a crane cable had broken while lifting a load of concrete into a multi-story hotel under construction, causing considerable damage to lower parts of the building. The crane company claimed the cable was defective since the load of concrete was well below the maximum load limit. The cable manufacturer claimed that the cable and crane operation allowed the cable to flex at the breakage area and in effect created a cyclic stress that was responsible for the underrated load failure.

The obvious analytic tact was to examine samples of the individual fractured wires composing the failed, braided cable, and especially to look for examples of fatigue stria-tions on the fracture surface using the SEM. In preparing samples for SEM fractography analysis, it was noted that the steel cable component wires exhibited two very distinct frac-ture features. Many wires appeared to be uniformly necked down, as in a regular tensile failure (see Figure 24a, for ex-ample, and compare with Figure 31a); others exhibited an irregular end form, as illustrated in Figure 31b. Closer inspec-tion in the SEM revealed about half the wires on one side of the cable to have failed in a normal, ductile-tensile mode, but the others were forcefully sheared and cut, probably by an ax.

It was obvious that the case involved criminal intent, and investigative authorities were notified. As it turned out, a disgruntled worker had sabotaged the operation by hacking about half the cable with an ax, reducing the load-bearing capacity by half and causing the remaining cable component wires to fail in service.

Case Example 6

In a case involving a wrongful death suit against a water tank manufacturer, it was claimed that a defective weld or welds in constructing the tank were responsible for its rupturing and killing a plumber.

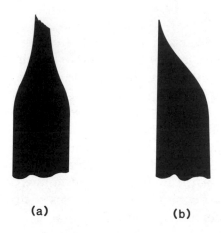

(a)   (b)

Figure 31.   Schematic representation of macroscopic fracture
morphologies observed for individual failed wires in a large braided
steel crane cable. (a) Symmetrical ductile tensile necking.
(b) Asymmetrically sheared or cut appearance.

The circumstances in this case were that a plumber had
been cleaning a filter system in a pump house on a ranch in
northern New Mexico. The pumping system included a deep,
submerged pump capable of providing a head pressure of over
300 psi. The 120-gallon water tank was a standard seam-
welded tank with convex top and concave bottom welded to
the cylindrical tank. The working pressure of the tank was
tested at 120 psi, and the tank was rated for 75 psi maximum
pressure in use. A tag on the tank indicated this maximum
working pressure.

Under normal operating conditions, this tank would
automatically relieve pressures in excess of 75 psi working
pressure, but the automatic relief valve could be bypassed by
a manual valve. During routine maintenance of the type the
plumber was conducting, the tank was to be valved off and
the submerged pump turned off.

During the maintenance visit, the tank ruptured by re-versing the concave bottom and ripping it along the circum-ferential weld seam, sending the tank 13 feet into the ceiling of the pump house, where it broke a hole through the roof and roofing tiles before falling back into the pump house and onto the head of the plumber, killing him instantly. In this "rocketlike" failure, pipes and electrical systems attached to the tank were ripped from the walls by the force of the ex-plosion, making it difficult to determine whether the system had been properly turned off.

During preliminary testimony in court, a state police officer called to investigate the accident, one of the first on the scene, testified that when he entered the sunken pump house standing water filled the entire floor area and covered the tops of his shoes. Calculations based upon a water depth of a few inches and considering the dimensions of the pump house clearly indicated a volume of water greater than 100 gallons.

The prosecution team and engineering consultants claimed the tank was properly serviced, was virtually empty, and an air pressure of a few psi was responsible for the adia-batic pressure failure of the tank bottom, which failed by improper welding practice. This contention was based on observations of large voids (some about 1 mm in diameter) in the fracture surface around the separated tank wall and bottom plate. Calculations of pressures necessary to cause the tank wall to fail allowing for 30 percent of the thickness to be lost to voids, however, indicated levels above 120 psi.

Field tests of similar tanks made in the same period clearly showed that rupture pressures were always in excess of 120 psi. SEM examination of selected specimens cut from the fractured bottom plate and tank wall clearly illustrated normal ductile-tensile fracture features even more pro-nounced than those shown in Figure 26 in the weld fracture, and it was concluded that the welds were technically sound.

Incidentally, in reinspecting the tank and attached pip-ing and electrical switches and connections, it was found that a manual bypass valve was in the open position. With this evidence, coupled with engineering tests, pressure-acceleration calculations of the height a failed tank might rise, and fractography and related fracture analysis data, the defense team contended that the plumber had failed to turn off the submerged pump, which could not be heard running from the pump house, and the tank filled with water and was pressurized well above the maximum sustainable internal pressure.

The prosecution brought in a welding specialist who correctly claimed that weld porosity is governed by engineering codes that prescribe techniques for measuring porosity against established standards. This welding-metallurgical specialist took X-ray shadowgraphs of the fractured welds around the failed tank circumference, using a technique similar to that illustrated in Figure 15. By measuring the average void diameters observable in the X-ray film, the void volume was calculated and shown to slightly exceed the allowable levels.

The defense countered that the weld porosity standards applied to as-welded products and that during fracture many intrinsic voids were created, as illustrated schematically in Figure 25a. It was argued that cracks created by void coalescence would be counted in the X-ray images and that even the propagation of cracks created additional voids and cracks that would continue to grow as the stress was increased to failure. To illustrate this, several dynamic sequences of crack growth and coalescence observed in a thin stainless steel film pulled in tension in a high-voltage transmission electron microscope were illustrated in court testimony. Figure 32 shows an example of such a sequence.

Although the illustration in Figure 32 is not directly related to the fracture of the water tank, it provides an example of fundamental materials behavior that can elucidate an important aspect of failure analysis. On the basis of this evidence and the supporting evidence alluded to previously, the porosity measurements and claims based upon the

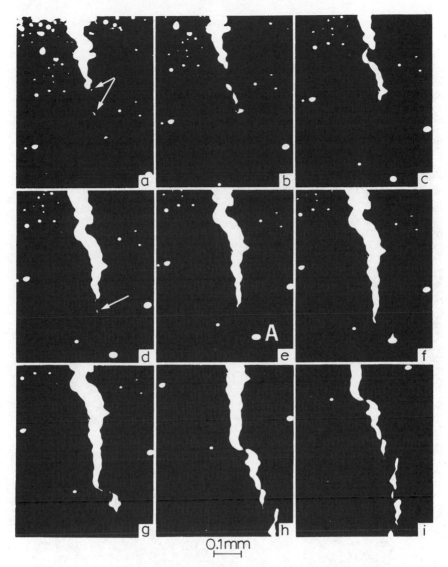

0.1mm

Figure 32.   Crack propagation sequence in a thin film of 304 stainless steel stretched in a high-voltage transmission electron microscope (observations made at 1 million volts). The arrows in the sequence a-d show microvoids and microcracks forming ahead of the principal crack tip. The field of view is moved as the crack advances from the film edge in a. A denotes the growth of a crack at a hole in the film. [From L. E. Murr, Structure and Properties of Tensile Cracks in Stainless Steel Films: In-situ, High-Voltage Electron Microscope Studies, *Int Journ. of Fracture, 20,* 117 (1982).]

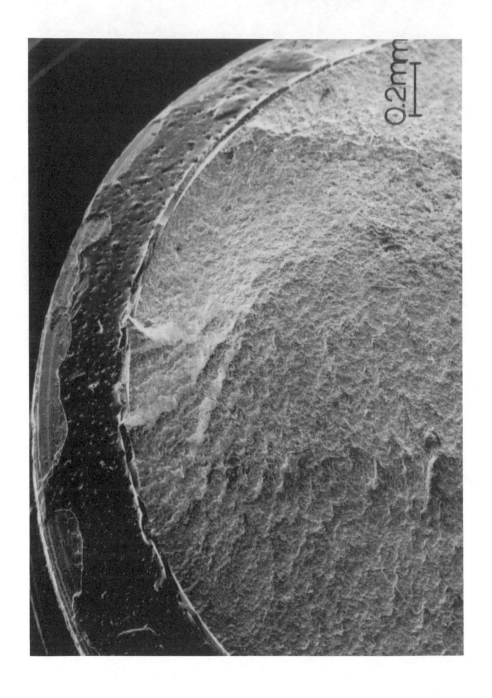

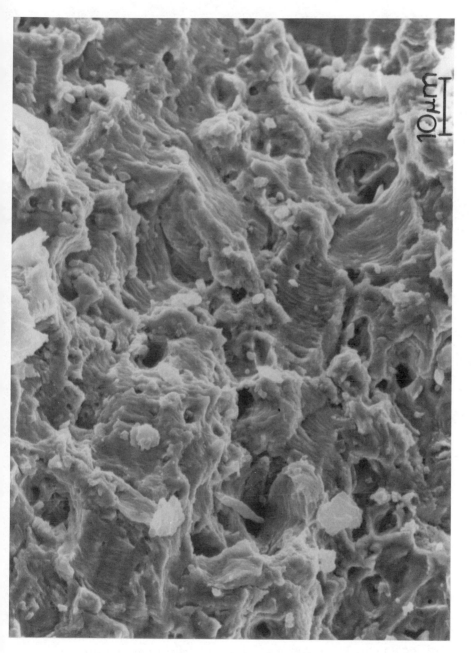

Figure 33. SEM fractographs showing fatigue failure in stainless steel prosthetic leg bone fracture implant screw. The lower magnification view shows a large portion of one fractured screw surface.

allusion to manufacturing standards were largely discounted by the jury. The jury in fact found no fault with the manufacturer.

Case Example 7

An attorney representing an elderly woman suffering repeated surgery for a leg bone implant problem requested an analysis of screws that failed, causing severe pain and requiring corrective surgery. The implant and screws were stainless steel and of a relatively standard design utilizing a total of six screws to attach the implant to stabilize an upper leg fracture and pull the fractured bone together.[14]

An SEM analysis of the failed screw fracture surfaces produced features shown typically in Figure 33 after the screws were ultrasonically cleaned in an ethyl alcohol solution to remove most of the body tissue and fluid contamination. These features show unmistakable fatigue striations (compare Figure 33 with Figure 29b, for example).

The prosthetic design did not and does not allow for screw fatigue. In the proper placement of the device, there should be no movement of the implant or cyclic loading of the screws. Because of the weight of the woman and her particular walking problem, she placed a cyclic load on two screws of the implant, and this was not properly included in the design.

Because the doctor was considered a specialist in this technique, and because it was demonstrated that he did not properly design the attachment and placement of screws, he was held liable for damages requested in a subsequent suit.

Case Example 8

As an additional example involving personal injury, we might consider a case involving a burn victim who filed suit against a camper and house trailer (recreation vehicle) manufacturer to recover medical expenses. In this case, the burn victim was visiting a relative and had parked the vehicle (an aluminum trailer pulled by a pickup truck) on the street in front of the relative's house. The victim and relative were talking in the trailer when the victim stepped into the small kitchen area to light the stove. An explosion occurred, throwing the relative and the victim from the trailer and blowing three sides off the trailer in the blast. The relative was miraculously un-injured, but the victim sustained severe burns over a large portion of his body. The heat of the blast was so severe that the aluminum structure was melted in many places.

When the explosion and fire were investigated by the fire marshall's office, the investigator noticed a break in a portion of the copper tubing connecting the propane gas sup-ply to the stove just below a flange connector. This was a complete break.

In the ensuing investigation, the victim was advised of the possible failure of the system, and he retained legal counsel to seek compensation.

As litigation continued, the failure of the gas line be-came a crucial issue. Initially the attorney for the burn victim believed that the gas line was somehow defective, it cracked, and gas leaked under the kitchen cabinets, which were relatively tight. Propane is a heavy gas so it will settle to the floor, displacing air as it fills space and rises. When the gas filled to the level of the stove burners, it ignited, causing the explosion.

The defense (attorneys for the distributor and manu-facturer) contended the pipe was broken by the force of the blast, and that other factors led to the explosion. The pipe failure, especially the mode of fracture, therefore became a crucial issue.

In analyzing the failed pipe section, the flange piece was placed in an SEM and the entire surface examined as shown in Figure 34. In addition to the flange fracture analysis shown summarized in Figure 34, a section of the tube was pulled to failure in tension and the fracture surface observed. These fracture surface features looked like those shown in Figure 26, which are expected in tensile-ductile fracture of

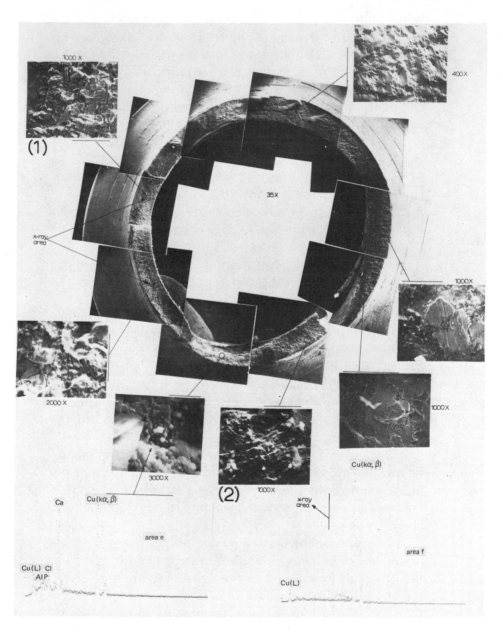

Figure 34.

copper tubing. An EDX analysis of the tubing and flange also revealed it to be common industrial copper tubing, and no chemical abnormalities were recorded.

It is apparent on examining Figure 34 that there is no evidence anywhere on the fractured flange surface of ductile-tension fracture. There is unmistakable evidence of fatigue failure, and several prominent examples are shown in magnified views in Figure 35. The strength of this argument is certainly reinforced on comparing Figure 35 with Figure 29b, which is also a similar copper product. In addition, the fracture surface shown in Figure 34 shows areas where the surfaces have rubbed together, obliterating any specific fracture surface features.

It was argued, on the basis of the fractography information, that the pipe could not have failed as a result of the explosion because it would have had clear tensile fracture features but no fatigue fracture striations. In addition, the design of the gas line demonstrated that a free-standing section under the kitchen cabinets would flex during road travel as the trailer rocked, concentrating a cyclic stress at the flange connector and the area where fracture occurred. It was therefore concluded that the pipe cracked by fatigue failure, allowing gas to leak into the cabinet space. The crack gradually grew, allowing the pipe to rub together, creating the

Figure 34.   Composite reconstruction of fractured flange end of copper gas line from SEM fractography views. The copper pipe size is standard 1/4 inch. Numerous strategic fracture site enlargements are shown, and faint reproductions of EDX spectra are included at the bottom of the composite. The spectrum at the bottom left shows calcium chloride crystals on the surface as a result of contamination or handling. The SEM enlargements at the upper right corner show obliteration of fracture features by the apparent rubbing of the tube surfaces. The X-ray areas indicated to the left exhibit EDX spectra identical to the copper spectrum of the lower right.

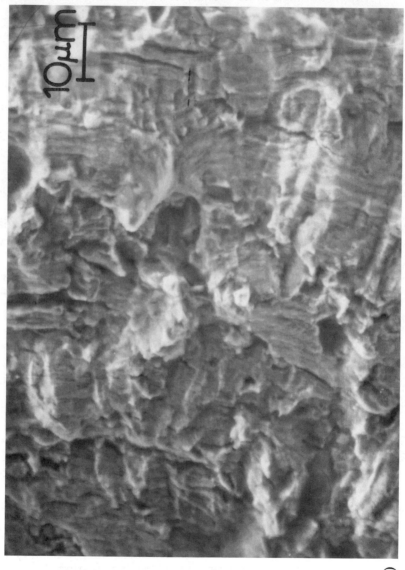

(a)

(b)

Figure 35. Enlarged views of SEM images marked (a) and (b), respectively, in Figure 34, showing fatigue striations superimposed on the fracture surface features. That fatigue striations occurred over a large area of the fracture surface is indicative of general catastrophic fatigue failure, and not localization of a fatigue phenomenon.

surface features observed in Figure 34. If the pipe had not completely failed prior to the explosion, it probably failed as a result of the blast, but a crack certainly existed prior to the blast.

With this evidence and the claim that the anchoring and design of the pipe probably contributed to fatigue failure (and thereby constituting an engineering design flaw in the manufacture of the vehicle), the manufacturer agreed to settle for damages out of court.

In these case examples (particularly Case Examples 7 and 8), the failure of a small component in a system (even a component within a human system, such as an implant) can be responsible for a very destructive effect upon the entire system. This is tantamount to a "weakest link principle" in which the system integrity is only as sound as its weakest link (or component). This is also a corollary of Murphy's law: *If anything can go wrong, it will.* Murphy's law is, in reality, one of the most important principles in the design process, which we will address briefly in the next chapter. It is important to recognize the sobering reality of Murphy's law and its engineering implications. In a complex materials system, Murphy's law implies that even the smallest component must be carefully designed and its relationship to the whole carefully considered. The space shuttle Challenger will forever remain a grim reminder not only of the weakest link principle and Murphy's law, but also of the necessity to carefully consider the behavior of materials in the overall design and manufacturing process.

Case Example 9

As a final example we will consider a case in which the failure of a small but crucial component in an automotive (truck) system led to a catastrophic accident. However, this is also an example in which the failure itself was related to another engineering issue. In addition, the analysis performed was not intended so much to determine the cause of failure as it was to provide engineering data needed to develop other conclusions relating to the accident or circumstances of the accident.

In this case, a truck (cab) was pulling a large house trailer. The trailer company involved was in fact transporting two trailers to a location, and the vehicle involved in this case was following a trailer hitched to another truck cab. On a fairly steep downgrade, the second "rig" lost its brakes and crashed into the lead rig and trailer, causing extensive damage to both at an intersection at which the first rig had stopped. The trailer company and the insurance carrier representing the company filed suit against the truck manufacturer, claiming that faulty manufacture of a brakeline connecting flange was responsible for the flange failure and the subsequent accident since the flange failure (a crack in a brake line connecting flange) allowed the leakage of brake fluid, rendering the braking system inoperable. Failure analysis in this case was intended in part to identify the brake line flange crack and its possible connection to manufacturing flaws in the material or process, but it was intended primarily to attempt to determine whether the brake fluid could have leaked in the downgrade application of the brakes just before the accident. This was important in part because the flange section cut from the brake line was a piece of evidence that could not be altered in any way (as specified by the attorneys involved), and as a consequence an engineering test of this piece of brake line in a pressurized system could not be performed. So the questions posed by the attorneys were: could the brake system have failed just prior to the accident, and could the fluid in the system have leaked so extensively during downgrade braking that the system failed completely?

In performing the analysis, it was important to characterize the crack in the flange. Figure 36 illustrates the flange section cut from the defective brake line portion (top) and a photographic enlargement showing the crack at the base of the flared end (bottom). Figure 37 shows further enlargements of crack sections (observed by optical photographic enlargements in Figure 36) observed by the scanning electron microscope. Examinations around the flare-end circumference indicated the crack to extend halfway around

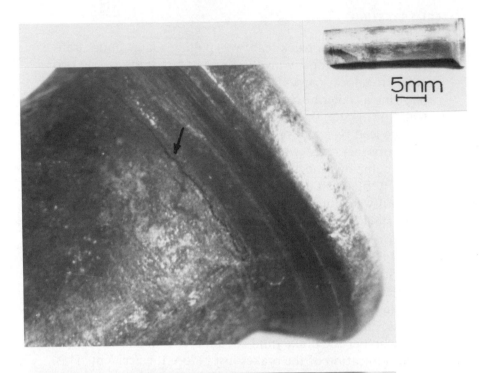

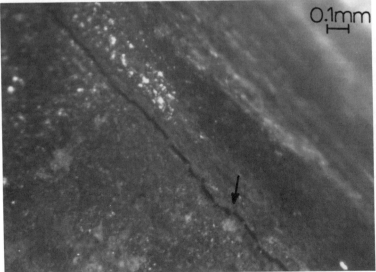

Figure 36. Optical metallographic views of cracked brakeline flange section. The arrow indicates a common point of reference.

0.01mm

Figure 37. SEM views of circumferential crack in brakeline flange of Figure 36. The magnification marker shown applies to the lower micrograph.

the tube and to penetrate completely through the tube wall. This latter feature was also confirmed by magnafluxing experiments.

The assumption was made that during actual service when the brakes were applied fully (maximum braking pressure), the crack in the hydraulic brake tube flare (Figures 36 and 37) would open, and the opening, or the increase in the crack width, could be estimated by estimating the maximum stress (and strain) in the tube wall. This strain can be calculated by knowing the internal pressure in the hydraulic brakeline and the elastic modulus of the tube. At most, it was estimated the crack would open or increase its width by 50 percent, but as a modest estimate it was assumed the crack width was to be represented throughout the wall thickness by that observed in Figures 36 and 37, and to extend roughly halfway around the tube circumference. It was then possible to estimate the rate of brake fluid loss from Bernoulli's theorem (illustrated in Figure 38), as follows:

Consider a more or less idealized case, as shown in Figure 38, in which a fluid escapes from a brakeline of cross-sectional area A_2. If the pressure in the line is P_1 (the hydraulic brakeline pressure), and the outside pressure (atmospheric pressure) is P_2, then the difference in pressure is given by

$$(P_1 - P_2) = \frac{\rho}{2}(V_2^2 - V_1^2)$$

where ρ is the fluid density, V_1 and V_2 are the fluid velocities (inside the brakeline and the velocity of the fluid as it escapes through the crack), and

$$V_1 A_1 = V_2 A_2$$

Since the outside pressure (atmospheric pressure) can be estimated to be 15 psi, and since the manufacturer's specifications for normal braking pressures P_1 were given as 1000 psi,

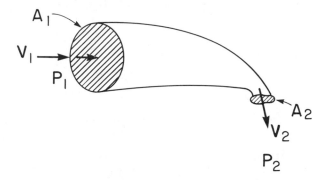

Figure 38.   Illustrating Bernoulli's theorem. A fluid in a system flowing with a velocity V_1, at pressure P_1, through a cross section A_1, is with a velocity V_2, at pressure P_2, through a cross section A_2.

the outside pressure is negligible by comparison and the following assumptions were made[15]:

$$P_2 \cong 0$$
$$P_1 - P_2 = \Delta P \cong P_1$$

Consequently, from the previous equations we can write

$$P_1 = \frac{\rho V_2^2}{2A_1^2} (A_1^2 - A_2^2)$$

This equation can be solved for the escape velocity of brake fluid through the crack:

$$V_2 = \sqrt{\frac{2P_1 A_1^2}{\rho(A_1^2 - A_2^2)}}$$

The rate of brake fluid loss can then be obtained from

$$\text{Rate of loss} = V_2 A_2 = A_1 A_2 \sqrt{\frac{2P_1}{\rho(A_1^2 - A_2^2)}}$$

In this case, remember that A_1 is the brake line tube (inside) cross-sectional area, equal to πr_1^2, where r_1 is the inside tube radius. A_1 was measured to be 0.0277 square inches. The crack width measured (estimated) from Figures 37 and 38 was 0.0007 inches, and the crack area A_2 can be estimated from

$$A_2 = \text{crack width} \times \text{crack length}$$

where the crack length is half the tube circumference, or approximately πr_1 (remembering that the circumference is $\pi \times$ diameter, or $2\pi r_1$, but only half the tube was cracked). Since the inside radius was measured to be 0.094 inches, the crack area A_2 was calculated to be 0.0002 square inches. Substituting these numbers, $A_1 = 0.0277$ square inches and $A_2 = 0.0002$ square inches, results in

$$\text{Rate of brake fluid loss} \cong 0.0003 \sqrt{\frac{P_1}{\rho}}$$

This equation is indicative that the rate of brake fluid loss is simply a function of the square root of the ratio of internal braking pressure to brake fluid density. We previously noted that the normal braking pressure was given as 1000 psi. For pressure in psi the units of brake fluid density must be slugs per cubic inch. For a nominal brake fluid density of 52 pounds per cubic foot (0.0316 pounds per cubic inch),

$$\rho(\text{brake fluid}) = 0.0001 \text{ slug/per cubic inch}$$

so the brake fluid would be lost at an estimated rate of

$$0.0003 \sqrt{\frac{1000}{0.0001}} = \sim 1 \text{ cubic inch per second}$$

Since the total capacity of fluid in the hydraulic brake system of the truck was 34.7 cubic inches, the brake fluid would be completely expelled in about half a minute.

Tubes in a test system were then cut with a fine saw to approximate the crack area opening, and the rate of brake fluid loss was actually measured. Measured rates varied from 22 to 28 seconds, but it was never possible to fully expel all fluid. Indeed, the system was inoperable (complete loss of braking action) when less than half the brake fluid was lost, leaving only roughly 15 seconds of braking action at the time of crack formation.

Several assumptions made in this analysis need to be discussed. First, it was assumed that the crack formed catastrophically just prior to the accident. If this were not the case, it could be argued that brake fluid would have leaked before the catastrophic accident, leaving even less braking time. Second, no account was taken of temperature effects on the brake fluid behavior. For example, data for the brake fluid showed that its viscosity R changed as follows with temperature:

$$R = 1800 \text{ centistokes at } -40°F$$
$$R = 4.2 \text{ centistokes at } 122°F$$
$$R = 1.5 \text{ centistokes at } 212°F$$

The actual fluid density is given by

$$\rho = \frac{R}{D}$$

where D is the diffusivity. So measurements made of fluid loss in laboratory tests at room temperature ($\sim 70°F$) probably underestimated the actual rate of loss because the brakeline was probably hot, reducing the fluid density considerably, and thereby increasing the rate of fluid loss.

These considerations and the loss rate, and other factors, were important in litigating the case because the manufacturer argued that the problem arose long after manufacture. During testimony at the trial, it was learned that the truck had been serviced just prior to the hauling exercise and the brake system checked. At this point, the plaintiff's attorneys began to question the tightening of the flange and the prospect of stressing it to failure.

In the final resolution of the case, negligence due to manufacturing or servicing could not be conclusively proven, and the insurance carrier was forced to cover the damages. During the ensuing arguments, however, it was pointed out that the flare design did lend to crack formation if overstressed by a mechanic putting too much torque on the connecting collar.

This example, in retrospect, points up a number of features. First, we attempted to demonstrate the use of failure analyis to gain quantitative engineering data. Second, simple engineering calculations were made based upon this data to provide guidelines in evaluating a part of the failure scenario. Third, the specter of gross variations imposed by temperature or other parameters was shown to have a potentially significant effect on very sensible engineering evaluations. Fourth, the implications for improving a poor design or in considering failure possibilities in the design of system components was touched upon briefly.

In the next chapter, we will examine these design implications a little more closely and develop some perspective regarding the prevention of component and system failures.

We will also examine the prospects of engineering assumptions and the dramatic effects that such factors such as temperature can have on the behavior of materials and the more complex, "real" issues that influence when and why materials fail.

5

Failure Prevention in Materials Systems: Engineering Design and Safety Factors

Failure prevention, even with the most effective design strategies and conservative use of simple engineering safety factors, can still pose a formidable task in a variety of materials systems. It is imperative to recognize that any materials system is only as stable or as dependable as its most unstable or least dependable component.

Any consideration of engineering materials and engineering design of systems utilizing materials must initially deal with Murphy's law for materials systems and the laws of materials applications:

Murphy's law: If any material can fail, it will.

Laws of materials applications

1. All materials are unstable.
2. The materials system is only as strong or as stable as its weakest or most unstable component.

   Although they are both obvious and incontrovertible, we might elaborate upon these laws, especially the laws of materials applications. All materials are indeed unstable. Even such materials as platinum can be degraded in particular environments. Under stress, all materials respond to that stress. Creep is an example. Given enough time, failure can be calculated to occur in all materials under creep stress. Stress and environment can act in concert to cause failure, for example, stress corrosion cracking. Creep-temperature effects and stress-temperature effects can contribute to the degradation of properties and components and system failure. The recent space shuttle Challenger tragedy (January 28, 1986) will serve as an example of the subtle effects that temperature can have on a system component's stability. Indeed, this is a classic example of the applicability of these laws.

THE DESIGN PROCESS

The design process usually begins with a specification of a solution. We sometimes allude to a design cycle, but the process may contain a design cycle plus design implementation, which involves actual production based upon the design. The design cycle can involve the original thoughts, sketches, and knowledge that in the specification stage produce engineering drawings. Computer-aided design[16] is now employed to implement a cycle in which various designs or design ideas may be tested or simulated.

Even in the early design cycle, it is useful to have some concept of available materials. Designs that cannot be achieved because materials with requisite properties do not exist are only concepts that will never be a reality unless a material or materials are developed. Prior to a fabrication step, materials specifications must be made depending upon the design requirements. These may have to be altered in the design simulation if the requisite properties cannot be identified in existing materials, or fabrication may have to be delayed until new materials can be designed and fabricated. Materials are the reality in a design.

Figure 39 illustrates a simple schematic representation of a design process that includes design implementation. The role of

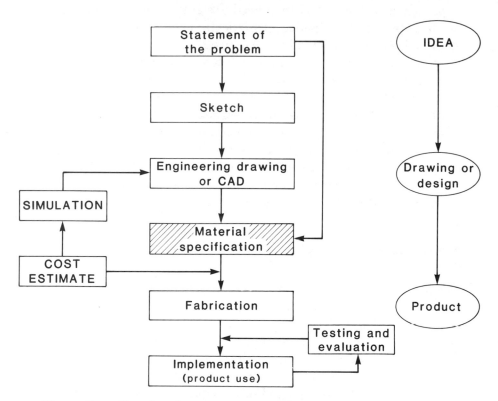

Figure 39. Simple schematic diagram showing a design and design implementation process and process cycles.

materials in the implementation is of course crucial because they make the design a reality. The design or design implementation diagram shown in Figure 39 may not characterize the final product. The first process may produce a prototype that will be tested and improved upon to develop a final product. Generally, a final product works reliably and is economical (or profitable) and safe.

SAFETY FACTORS IN ENGINEERING DESIGNS

The concept of safety factors is normally applied to the evaluation and consideration of strength of materials used in a design. Safety factors are normally applied to the yield strength of a material but can be applied to the ultimate tensile strength. If we look in retrospect at Figure 11, the yield stress is the stress above which the material (particularly a metal or alloy) will deform permanently. Having exceeded the yield stress or yield strength, materials begin to creep. That is, dislocations will be created, and application of a load to continue the stress above yield will allow the material to slip. Over time, it may slip enough to simply come apart. So if a material is used in a service environment in which the yield stress is never exceeded, it is unlikely it will fail by being overstressed. We sometimes base a design on fractions of the yield stress. For example, if we assume a material will never exceed half the value of the yield stress and use this in the actual design, we can realize a safety factor of 2. In an extreme case, we might allow the maximum stress in the system to be $\sigma_y/4$, and the safety factor would be 4. It can be observed in Figure 11 that even a safety factor of 2 based upon the ultimate tensile stress would still be below the yield stress σ_y. Obviously a design that would put a material into a service environment in which the normal stresses would be near the ultimate tensile stress for sustained periods of time would be unsafe. On the other hand, large safety factors could render a design unworkable because no materials could be identified with the requisite strengths. In addition, even fulfilling the strength requirements does not necessarily justify a material's use. The material may be too expensive, or too corrosive, or unstable in other ways for the intended application. Or perhaps, in addition to high strength, high electrical conductivity would be required.

REAL SYSTEMS, REAL PROPERTIES, AND THE REAL WORLD

In describing some of the simple concepts of engineering materials safety factors above, it was tacitly assumed that we were only considering specifications of strength based upon a uniaxial tensile stress-strain diagram (Figure 11). In such a situation, design specifications are based simply upon Hooke's law (Figure 11):

$$\sigma = E\epsilon$$

corresponding to a single temperature or a limited range of temperatures and some specific strain rate.

   In reality (in a real system), this simple assumption or simple design specification may be totally inadequate, and even safety factors based upon this assumption would not assure a successful product, that is, one that would not fail. Even for a uniaxial (tensile) stress system (referred to as a uniaxial stress state), changes in plastic stress are ideally expressed by a mechanical equation of state (or a modeling expression) of the form

$$d\sigma = \left(\frac{\delta\sigma}{\delta\epsilon} \right)_{\dot{\epsilon},T} d\epsilon + \left(\frac{\delta\sigma}{\delta\dot{\epsilon}} \right)_{\dot{\epsilon},T} d\dot{\epsilon} + \left(\frac{\delta\sigma}{\delta T} \right)_{\epsilon,\dot{\epsilon}} dT$$

   For those with limited experience with or understanding of differential equations, this may appear to be a rather formidable and ominous mathematical expression. Symbolically, however, it is a statement of the fact that changes in plastic stress, even in the case of simple tension (stretching of a metal wire, for example), are really dependent upon the strain ϵ, the rate of straining $\dot{\epsilon}$, where

$$\dot{\epsilon} = \left(\frac{d\epsilon}{dt} \right)$$

or change in straining with time at temperature T. We might write this as a simple functional relationship:

$$\left[\Delta\sigma = f(\epsilon, \dot{\epsilon}, T) \right]_I$$

where the Roman numeral I refers to a uniaxial strain state (stretching a wire). So in reality, even stretching a metal wire is better represented by a multiparameter plot, as shown, for example, in Figure 40 for mild steel. The reader must realize that Figure 40 characterizes changes of stress σ and strain ϵ. It represents only a single temperature, however. Over a range of temperatures, multiparameter plots like Figure 40 might change dramatically.

Now if Figure 40 seems complex in comparison with Figure 11, for example, in which strain rate is not shown as a parameter, you, the reader, must recognize that the functional relationships for mild steel shown in Figure 40 will probably not describe mild steel subjected to stress or strain in two directions (biaxial). As you might already have concluded, the same mild steel subjected to triaxial stresses will be described by still different parametric relationships. So if we wanted to provide some idea of how complicated it might be to describe how a single material might behave in each of three stress states (uniaxial I, biaxial II, or triaxial III, we might have the following:

$$[\sigma = f(\overset{\downarrow}{\epsilon}, \dot{\epsilon}, T)]_I$$
$$[\sigma = g(\epsilon, \dot{\epsilon}, T)]_{II}$$
$$[\sigma = h(\epsilon, \dot{\epsilon}, T)]_{III}$$

where f, g, and h represent different functions. The arrow in this block of equations is indicative of the fact that if a design

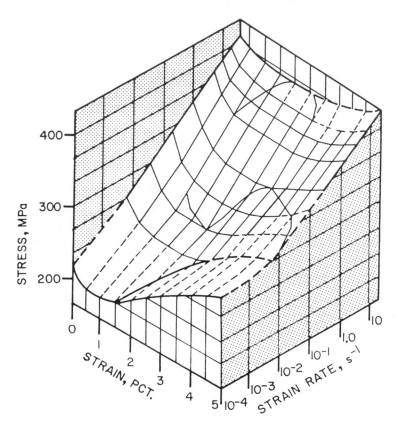

Figure 40.   A three-dimensional plot of stress σ, strain ϵ, and strain rate $\dot{\epsilon}$ for a mild steel at room temperature. (After J. D. Campbell, *Dynamic Plasticity of Metals*, Springer-Verlag, Vienna, Austria, 1972.)

specification is based upon a stress-strain diagram (Figure 11) at some specific temperature and strain rate, and considering simple tension, it could be risky at best to use the design at another temperature or strain rate, even within the same stress state, let alone biaxial or triaxial stress states.

Let us look at this whole issue with a few simple examples in which temperature is a parameter. Figure 41a shows some

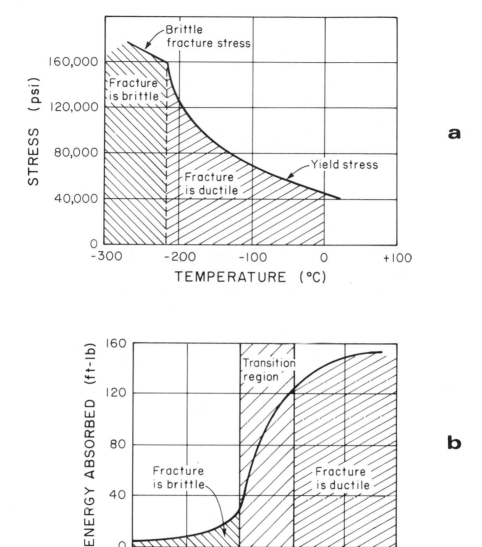

Figure 41. The ductile-brittle transition in low-carbon steel.
(a) Tensile test results based upon data of A. S. Edlin and S. C.
Collin *(Journal of Applied Physics, 22, 1296, 1951)*. (b) Impact
test results based upon data of G. C. Smith (*Advances in Materials*,
P. A. Rottenbury (ed.), Pergamon Press, New York, 1964).

results for tensile tests of a low-carbon steel over a range of temperatures at low strain rates (10^{-4} per second). Above about -220°C, this material is reasonably ductile, and as temperature is lowered the yield stress actually increases significantly. So in effect, if this material were in service at room temperature (~ 20°C) the yield stress would be 40,000 psi, but at -200°C, the yield stress would be three times greater. The yield stress effectively rises even below -220°C, although the material becomes brittle. This means that when the yield stress is reached, it will not stretch further but will simply break.

Now compare this behavior at low strain rate with the behavior under impact loading (at strain rates possibly in the range of 10^2 per second or a million times faster than at the low rate in Figure 41a. Figure 41b shows the same low-carbon steel tested by impacting notched samples at different temperatures (see Figure 24). There is still a brittle-ductile transition, but the transition is more gradual than in Figure 41a. Moreover, the transition is shifted in temperature (raised) by some 200°C. This occurs because of the impact loading compared with low strain rate tensile loading. Figure 41b also shows the very dramatic behavior of low-carbon steel under impact loads. At room temperature (~ 20°C), the material can sustain a large impact blow. Below -20°C, the material would fracture in a brittle fashion with a load about 200 times smaller. This dramatic difference in brittle-ductile behavior occurs in a range from room temperature to freezing. Here again we can invoke the example of the space shuttle tragedy. Materials can catastrophically alter their behavior over a narrow temperature range and in response to dramatically different loading. Plastics and other polymers (including rubber) behave in a similar and often more dramatic way.

THE MATERIALS SYSTEM

Finally, let's look at the concept of a materials system. A system in the most general sense involves an arragement of and connection of parts or elements into a whole. A machine is therefore a mechanical system having connected components. A materials system is therefore an arrangement of connected materials. The system can be partitioned into different materials where they connect, and in the most ideal system materials are separated by an interface or boundary that connects one component to another or provides a transition from one regime to another. Fracture, as we have discussed [in Chap. 4; e.g., Figure 29a], can occur at crystal interfaces, and therefore even a polycrystalline metal or alloy might be thought of as a system of connected crystals. The dispersion of ceramic particles in a metal matrix (as illustrated in Figure 6a), composites of glass fibers in a plastic binder, or inorganic crystal needles in a metal matrix (metal-matrix composites) are also materials systems.

Figure 42 illustrates a rather idealized view of a multilayer, multiphase, multicomponent system. This system is characterized by seven regimes (phases) that include the air surrounding the system (phases I and VII). There are three fundamental states or phases of matter: gas G, liquid L, and solid S. The interface between the solid and gas (air), which we normally regard as the "surface," is a gas/solid interface (S/G). The interfaces separating the solid phases are solid/solid (S/S) interfaces. In the example shown in Figure 42, metals, ceramics, polymers, and semiconductors (Figure 1) can be connected in arrangements to produce a solar reflector or a solar cell (or photovoltaic device). Modern integrated circuits or VLSI (very-large-scale integrated) circuits can also be modeled ideally by a system like Figure 42 involving connections or layers of metals, ceramics (oxide insulation layers) semiconductors, and polymers (plastics as hermetic seals, and so on).

In the context of failure as we have discussed it throughout this book, it is important to realize that failure cannot only occur within a regime (a component in a connected arrangement), but at the connection itself. Intergranular fracture, as shown in Figure 29a, occurs by decohesion at the crystal interfaces. So the interface is also a part of the system and must be considered in the context of the weakest link corollary and Murphy's law.

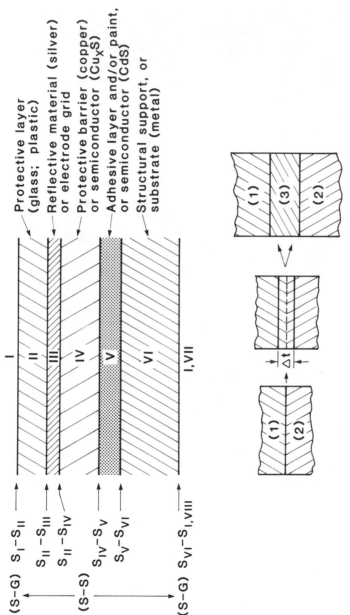

Figure 42. Idealized interfaces and interfacial phenomena in materials systems. The upper construction can be considered to schematically resemble a typical solar reflector or thin film solar cell (photovoltaic device) showing multilayer, multiphase, multicomponent contributions to forming solid/solid interfaces. Systems are denoted I–VII where I = VII = AIR. A total of six interfaces are shown. In the lower schematic an interface of finite thickness Δt reacts to become a new phase (3). [From L. E. Murr, *Materials Science and Engineering*, Vol. 53, p. 25 (1982).]

As illustrated in Figure 42, interfaces can be regarded as a finite phase region. A grain boundary or separation between two crystals in a polycrystalline metal or alloy is a unique regime that may be regarded as having a thickness of only a few atomic dimensions. Nevertheless, it is a unique component and will, under certain circumstances, behave quite differently from the crystal matrix. As illustrated in Figure 42, an interface phase may form as a result of a chemical reaction between the two phases separated by the interface, either by selected diffusion of chemical species to the interface from the separate phases, or by the intrusion of another chemical species along the interface from the environment in which the system is operating. When this occurs, the system itself is altered and a new component with different and unaccounted-for behavior may be introduced.

Figure 43 illustrates an example of a new component phase forming in an electronic system as a result of atmospheric moisture facilitating a reaction between an aluminum connector on a silicon semiconductor base. The reaction product, having the appearance of my grandmother's armchair doily, is a nonconducting aluminum silicide. When this problem occurred in the late 1960s in thousands of imported pocket radios in humid cities like Houston, Texas, it was a classic example of unexpected system failure by a component that never existed in the original system design. Using unreactive electrodes eliminated the problem which was readily solved through such observations as that illustrated in Figure 43 (another classic example of Murphy's law).

Figure 42 can serve as an example of how complex a materials system can be. Failure may include mechanical or electrical failure (as implicit in Figure 43), or both, and the creation of a new phase or environmental or chemical alteration of an existing system component may also contribute to failure. The suspected alteration of an existing system component may also contribute to failure. The suspected alteration of properties of rubber seals joining rocket segments in the space shuttle Challenger is perhaps the epitomy of this phenomenon.

Failure prevention is certainly not a hopeless issue even in very complex materials systems but it does require a great deal of

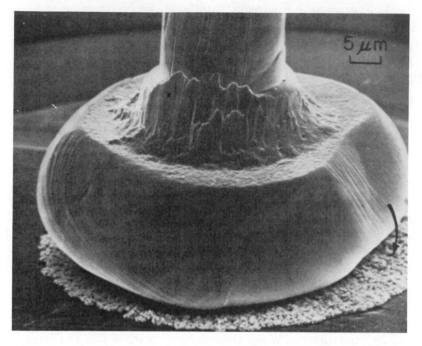

Figure 43.   SEM view of reaction product (aluminum silicide) formed in older electronic pocket radio by a reaction between the aluminum contact and the silicon base in humid air. The reaction product (arrow) is nonconducting and acts like a sheet of paper stuck between the electrical contact and the silicon device. This occurrence was referred to as "purple plague." (From L. E. Murr, *Solid State Electronics*, Marcel Dekker, Inc., New York, 1978.)

testing, evaluation, and fundamental understanding of materials behavior under a variety of conditions. Selecting optimum conditions, underutilizing components in a system (designing for the weakest link and employing safety factors), introducing safety factors, consideration for environmental changes and even bizarre conditions of use — all contribute to successful applications of products. One need only to take a glimpse at the technology around us to appreciate how successful we have been in designing efficient, dependable, and useful products and fixtures.

6

Responsibility, Liability, and Litigation

Liability in its simplest terms is a condition of being responsible. Litigation is the act of carrying out a suit or claim in a court of law; litigation involves the process of deciding liability.

In a companion book in this series, Professors James F. Thorpe and William H. Middendorf, in their introduction to *What Every Engineer Should Know About Product Liability*, write the following:

> The relationship between manufacturer and user and the design atmosphere have changed to the spirit of caveat venditor (let the seller beware). Today the designer must not only be able to apply the technical elements of design but must also meet society's expectation of a design. During the design process itself this means using far broader design criteria, including failure analysis, fail-safe design and redundancy, product safety audits, hazard analysis, and compliance with government and voluntary standards. In managing for design process this means an expanded array of *responsibilities*, including design review, product liability prevention, and documentation. . . .
>
> It is the engineer, in this milieu, who has a unique opportunity to play a professional role. Whether prospectively in product design, retrospectively in litigation, or as an educator training the next generation of professionals, the major *responsibilities* for the technical decisions that can provide reasonable solutions to the demands for product safety are those of the engineer.

Engineers are therefore responsible at the very least to adhere to a code of professional ethics. The following statements, prepared by the Ethics Committee of the older ECPD (Engineer's Council for Professional Development), the predecessor of ABET (Accreditation Board for Engineering and Technology), continue to provide a simple, if not idealistic, framework:

FAITH OF THE ENGINEER

I am an engineer. In my profession I take deep pride, but without vain-glory; to it I owe solemn obligations that I am eager to fulfill.

As an Engineer, I will participate in none but honest enterprise. To him that has engaged my services, as employer or client, I will give the utmost of performance and fidelity.

When needed, my skill and knowledge shall be given without reservation for the public good. From special capacity springs the obligation to use it well in the service of humanity; and I accept the challenge that this implies.

Jealous of the high repute of my calling, I will strive to protect the interests and the good name of any engineer that I know to be deserving; but I will not shrink, should duty dictate, from disclosing the truth regarding anyone that, by unscrupulous act, has shown himself unworthy of the profession.

Since the Age of Stone, human progress has been conditioned by the genius of my professional forbears. By them have been rendered usable to mankind Nature's vast resources of material and energy. By them have been vitalized and turned to practical account the principles of science and revelations of technology. Except for this heritage of accumulated experience, my efforts would be feeble. I dedicate myself to the dissemination of engineering knowledge, and, especially to the instruction of younger members of my profession in all its arts and traditions.

To my fellows I pledge, in the same full measure I ask of them, integrity and fair dealing, tolerance and respect, and devotion to the standards and the dignity of our profession; with the consciousness, always, that our special expertness carries with it the obligation to serve humanity with complete sincerity.

It is indeed somewhat reassuring to realize that essentially every engineer and every engineering manager connected with the failed rocket booster of the space shuttle Challenger made recommendations against the fateful flight or expressed concerns about the safety and reliability of the space shuttle booster system as a result of problems that were evident during design and testing.

LIABILITY CONCEPTS

Liability in its simplest terms is a condition of being *responsible* for an actual or possible loss, penalty, or burden. Thorpe and Middendorf,[17] in their treatment of liability, describe contractual and tort concepts. In the former, a contract, warranty, or guarantee can represent an agreement that is *expressed* (usually in writing). Failure to comply with this agreement or a violation of this agreement constitutes a liability. There is also an *implied* warranty that carries the expectation that any product or component offered for sale is "reasonably" safe. In the latter concept, a tort represents a wrongful or "irresponsible" act that results in injury (even to a successful business, for example) and from which some civil legal action (litigation) may result. To establish liability in the tort concept, it is therefore necessary to establish negligence (when the standard is based on what a "reasonable" engineer, designer, or manufacturer would have done), which is referred to as absolute liability, or to demonstrate strict liability. In strict liability, it is not necessary to demonstrate negligence on the part of the manufacturer, engineer, or designer but simply to demonstrate that a product or component was defective, unreasonably dangerous, and the proximate cause of injury.

Indeed, the reader should recognize in retrospect that this was the theme of nearly all of the Case histories (Cases 2-9, in fact) presented previously in this book.

In 1965, there were roughly 100,000 liability suits in the United States. By 1985, this number is estimated to have been in excess of 2.5 million product liability suits. As an example of the awards in such litigation, American Network, Inc., of Vancouver, Washington, won a $37.5 million jury verdict in March 1986 against a Texas telecommunications equipment company that Amnet allegedly sold it malfunctioning telephone switching machines. This has created a crisis in insurance for many companies and professionals. The problem does not stop with products, but has now been extended to medical professionals and a host of others providing services without products. In an article written by Donald Peckner in the September 1977 issue of *Metal Progress*, he begins, "Five years ago this article would probably not have been written. Although the likelihood that the individual engineer would be found liable in a product case is still questionable, recent trends in the law have made this a real threat." Indeed, this threat is now a demonstrated reality in many court records.

As a matter of professional responsibility and as a means to reduce the risk of liability, it is incumbent upon every engineer to

Conform to the highest standards of his or her field; to "keep the faith" (see p. 123).

Follow all safety requirements, design codes, standards, and practices in a responsible fashion. Don't take reckless shortcuts, and don't be pressured by anyone into abandoning your professional principles. Have the guts to express a concern for safety even if you may be wrong.

If compromises must be made in product design or development, *clearly* state the limitations of use. Liability is often demonstrated simply as a failure to reasonably inform a user of a product's limitations.

Take Murphy's law seriously. Consider potential design flaws or materials failure in light of a careless or reckless user.

Try to make the product as foolproof as possible. Be concerned about a "worst-case scenario."

LITIGATION: A SUMMARY AND CLOSURE

Litigation is the act of carrying on a suit or claim in a court of law; a judicial contest involving any controversy that must be decided upon evidence. Litigation involves the process of *deciding liability*. In the case of product failure or material or component failure, failure analysis provides evidence. Engineers actually provide such evidence; attorneys present the evidence and argue for or against liability on the basis of how they interpret the evidence.

To a large extent, this book has been about litigation. Case histories have illustrated the litigation process not only from an engineering perspective but also from a legal perspective as well. Litigation has been used in this book to illustrate responsibilities and liabilities of an engineer in the design, manufacture, and testing of materials systems.

Although it has not been possible or practical to provide a complete overview of materials properties and the materials sciences, it is hoped the reader will have gained some real appreciation for the role that materials science and engineering play in structuring our world and the products of our technologies. It is also hoped that the engineer and engineering student and lawyer or law student who might read this book will have gained some real insight and hindsight into the examination and analysis of material and component failure, and advantages and disadvantages of sophisticated analytic techniques and some basis for understanding the complex behavior of materials, particularly metals and alloys.

Notes

1. The following materials science and engineering texts will provide additional background and principles: L. H. Van Vlack, *Materials Science for Engineers*, Addison-Wesley Publishing Co., Reading, Massachusetts, 1970; H. W. Pollack, *Materials Science and Metallurgy*, 2nd edition, Reston Publishing Co., Reston, Virginia, 1977; C. S. Barrett and T. B. Massalski, *Structures of Metals*, 3rd edition, McGraw-Hill Book Co., New York, 1966; A. G. Guy, *Essentials of Materials Science*, McGraw-Hill Book Co., New York, 1976; J. E. Gordon, *The New Science of Strong Materials or Why You Don't Fall Through the Floor*, Princeton University Press, Princeton, New Jersey, 1984; J. E. Neely, *Practical Metallurgy and Materials of Industry*, Wiley, New York, 1984.

2. See L. E. Murr, *Electron and Ion Microscopy and Microanalysis: Principles and Applications*, Marcel Dekker, Inc., New York, 1982.

3. See L. E. Murr, *Interfacial Phenomena in Metals and Alloys*, Addison-Wesley Publishing Co., Reading, Massachusetts, 1975, for a detailed account of interfacial phenomena.

4. See P. Gordon, *Principles of Phase Diagrams in Materials Systems*, McGraw-Hill Book Co., Inc., New York, 1968.

5. See the historical perspective provided by L. S. Darken and R. W. Gurry in *Physical Chemistry of Metals*, McGraw-Hill Book Co., New York, 1953. This reference also discusses the Fe-C system. A comprehensive overview of iron and steel is given in *The Making, Shaping, and Treating of Steels*, edited by H. E. McGannon, United States Steel Corp., 9th edition, 1971, or more recent editions.

6. For an introductory treatment of dislocations, dislocation interactions, and applications in materials, the reader might peruse D. Hull and D. J. Bacon, *Introduction to Dislocations*, 3rd edition, Pergamon Press, Oxford, 1984.

7. R. C. Gifkins, *Optical Microscopy of Metals*, Elsevier, New York, 1970.

8. See *Metallography in Failure Analysis*, J. L. McCall and P. M. French (eds.), Plenum Publishing Corp., New York, 1981.

9. See R. V. Williams, *Acoustic Emission*, Hayden Press, Philadelphia, 1980. J. M. R. Weaver, C. Ilett, M. G. Somekh, and G. A. D. Briggs, Acoustic Microscopy of Solid Materials, *Metallography*, *17(3)*, 3 (1985).

10. C. F. Quate, The Acoustic Microscope, *Scientific American*, *24(4)*, 58 (1979).

11. C. F. Quate, Acoustics of Microwave Frequencies for Thin Film Studies, *Thin Solid Films*, *84(2)*, 184 (1981).

12. A. Taylor, *X-Ray Metallography*, J. Wiley, New York, 1961.

13. See Metal Handbook, Vol. 9, 8th Edition, *Fractography and Atlas of Fractographs*, American Society for Metals, Metals Park, Ohio, 1974.

14. The reader should consult the reference in Note 8 for a detailed description of such implants and related fractography and metallography applications.

15. The units in this case example remain the same as those developed in the initial analysis. Juries and litigants usually

have difficulty understanding the SI units or the metric system in general. This is important to realize in descriptions of engineering evaluations and failure analysis evidence.

16. The reader might find the companion book in this series helpful, *What Every Engineer Should Know About Computer-Aided Design and Computer-Aided Manufacturing* by John K. Krouse, Marcel Dekker, Inc., New York, 1982.

17. The reader should refer to *What Every Engineer Should Know About Product Liability* by James F. Thorpe and William H. Middendorf, Marcel Dekker, Inc., New York, 1979.

Glossary

Alloys Combinations of elements forming metals. For example, steels are alloys because they are combinations of iron and carbon or iron, carbon, and other elements. Brass is a combination of copper and zinc, and it is an alloy.

Amorphous materials Materials that do not possess any recognizable or definable crystalline structure.

Analytical electron microscope An electron microscope, usually a transmission electron microscope, that allows the electron beam to be scanned across an area of a very thin specimen of material, which is fitted with detectors for characteristic X-rays or for differentiating the energies lost by electrons passing through the specimen, which can allow a chemical or other analysis to be made for the thin sample material.

Atomic number The number of protons in an atomic nucleus, consequently referred to at times as the proton number. This is normally designated by Z.

Atomic structure   The arrangement of the components of an atom, including a nucleus containing positively charged protons and neutrons and a complex arrangement of negatively charged electrons surrounding the nucleus with the number of electrons equal to the number of protons. The atomic structure is most apparent, however, by examining its electronic structure; sometimes atomic structure and electronic structure are used interchangeably.

Auger electrons   Sometimes called characteristic electrons, Auger electrons are electrons ejected from specific energy levels for particular atoms within a solid volume bombarded by some energetic radiation or particle beam. These electrons therefore possess unique energies characteristic of the atoms from which they were ejected.

Auger spectrometer   Solids irradiated by energetic beams of electrons, ions, or X-rays, for example, can eject electrons from specific electron shells or energy levels. These specific or characteristic electrons are called Auger electrons. A device that can separate these electrons and identify them according to their particular energies and then identify the atom and electronic shell from which the Auger electron was ejected is an Auger spectrometer. This separation process is accomplished by directing the Auger electrons through a magnetic field that causes the Auger electrons to describe different paths depending upon their energy.

Austenitic phase   Iron can arrange itself into a number of crystalline structures. The face-centered cubic structure is the gamma (γ) phase, or austenite.

Backscattered electrons   When a primary beam of electrons strikes a surface, particularly a solid surface, a certain fraction of these electrons will simply scatter or bounce back from the surface. These are said to be backscattered.

Biaxial loading   Two principal stresses (corresponding to directional axes x and y, for example) operate on any point in a body. Forces acting in two directions resolved to be perpendicular to one another.

Boule A pure, continuous crystal formed synthetically by growing it from a melt by rotating a small seed crystal while pulling it slowly out of the molten material. Such boules are usually cylindrical and can be 8 inches in diameter and 4-8 feet in length for materials, like silicon, grown to fabricate wafers for semiconductor device substrates.

Brittle fracture Abrupt decohension of a solid without any noticeable plastic deformation or plastic strain. In brittle fracture a material will not stretch but simply break abruptly when some critical force or stress is reached.

Carbon steels Most steels contain carbon but alloys composed of only iron and carbon are called carbon steels, as opposed to alloy steels, which can, in addition to iron or iron and carbon, contain a variety of other elements.

Cast iron Any iron-carbon alloy that contains 1.8 - 4.5 percent carbon cast to shape.

Cathode-ray tube See CRT

Characteristic x-rays When a solid is irradiated by energetic beams, such as electrons, the atoms composing the solid in the irradiated area are excited and electrons can be knocked out of specific electron energy levels. When this occurs, electrons in higher levels or shells composing the atoms will fall down into the lower shell from which an electron has been ejected. When it does this it will lose energy, and this energy loss can occur as an X-ray photon. Since this energy loss or the energy of the x-ray photon released will be specific to each such process for each atom, the X-rays produced are said to be characteristic of that atom, or characteristic X-rays.

Computer-aided design (CAD) Designs can be simulated and tested using CRT display features in connection with a computer. This can be an aid in the design process, but entire designs can also be developed.

Creep Creep is the flow or displacement of a material over a period of time when under a load (force) or stress too small to produce any measurable plastic deformation at the time of its application. The stress applied over time is therefore below the yield stress.

CRT Abbreviation for cathode-ray tube. An electron tube or vacuum tube in which a beam of electrons can be focused to a small spot or area on a fluorescent screen, which can be varied in position and intensity on the fluorescent screen. A television tube is a CRT, as is the video display for most computer terminals. In these displays the focused beam is rastered across horizontal lines on the screen to develop an image.

Crystal grains See Grain structure

Crystalline arrangements See Crystalline structures

Crystalline structures Unique, periodic spatial arrangements of elements or atoms composing a solid.

Crystallization The formation of crystalline substances from solutions or melts.

Cyclic stress A stress applied so that a maximum is reached and the stress direction is reversed or the stress reduced and then increased again. A fatigue stress is a cyclic stress that can alternate between tension and compression, a positive and negative bending moment, or zero and some value in repeated cycles over time.

Deformation Any alteration of shape or dimensions of a body as a result of applied stresses, including thermal expansion, applied mechanical forces, thermal contraction, or other chemical or metallurgical transformations creating internal stresses.

Diffusivity The relative movement of a particle in a system (including a gas, liquid, or solid system) relative to its neighbors is measured by a flux or transport of particles per unit time per unit area. The diffusivity or diffusion coefficient is proportional to the particle flux or transport and is measured in unit area per unit time.

Dispersion hardening Solid materials can be hardened by dispersing inclusions or creating precipitates of a different phase within the solid. These particles impede deformation and consequently contribute to its hardness.

Ductility The ability of a material to be plastically deformed without breaking. Deformation is usually measured by elongation for materials deformed in simple tension tests. More ductile materials will therefore elongate more in proportion to an applied stress.

Electromagnetic spectrum The total range of frequencies or wavelengths that characterize electromagnetic radiation. This spectrum of course contains radio waves, light waves (including the range of colors composing that portion of the spectrum), X-rays, and cosmic rays. Arranged according to increasing wavelength, each of these radiation types or waves assumes a specific position in the spectrum.

Electron diffraction The scattering of electrons or a beam of electrons from matter. In crystals the scattering is systematic. The scattered electron intensity can vary because of interference effects, which can also be systematic.

Electron microscope See Transmission or Scanning electron microscope.

Electronic structure The structure of atoms or molecules is ultimately differentiated by differences in the number and arrangement of the electrons composing the atoms or molecules.

Electrostatic lens An arrangement of electrodes to create electrostatic fields, which can be shaped by the electrode arrangement to act upon charged particles or particle beams similar to the way in which a glass lens acts on light beams.

Equation of state Any equation that contains all the parameters or information about a given substance or material in a particular context is said to describe the state of the material in that context. For example, an equation can contain all the thermodynamic information of a material or the mechanical information.

Eutectic An alloy or solution that has the lowest possible melting point. For example, mixing two elements A and B and noting the melting points for all possible combinations may show a unique composition of A and B for which the melting point is lowest. This is the eutectic. When an alloy having this composition solidifies, it will have a unique microstructure as well.

Failure analysis An examination of the characteristics and causes of failure. In materials failure analysis this can involve fractography, chemical analysis of the failed members, an examination of the system design or engineering design, the applications and uses of failed members, and so on.

Fatigue (metal) The failure of a metal or other materials or materials system under prolonged, repeated, or cyclic stress. Fatigue occurs at much lower stresses than tensile failure. For example, a thin metal wire can be bent back and forth to failure, but a great deal of strength might be required to pull the wire apart.

Ferrite phase Iron can arrange itself into a number of crystal structures. The body-centered cubic structure is called the alpha (a) phase, or ferrite.

Field-ion microscope A special microscope in which the surface atoms composing a tiny whisker or metal needle can be made visible by ionizing a gas in the viewing chamber by applying a high voltage between the needle surface and a fluorescent screen. The charged ions follow electric field lines emanating from each surface atom and create an illuminated spot on the screen. The whisker or needle to be viewed must be cooled to several hundred degrees below zero Fahrenheit to minimize the vibrations of the atoms in order to see distinct images. Because of the projection geometry from the needle surface to the screen, the magnification of surface atomic structure is enormous, usually a million times or more.

Fractography The microscopic examination of the structures or microstructures on the surfaces of fractured materials, particularly metals and alloys.

Fracture modes Fracture can occur in different ways or modes depending upon the nature of the material and the applied stress. For example, fracture can be brittle or ductile, or it can occur by fatigue through the application of a cyclic stress. Fracture might also occur by impact or shock stresses.

Grain boundary The region or interface that separates one crystal from another in a polycrystalline solid. Generally these crystals will have the same structure and composition, but not necessarily.

Grain boundary segregation Depending upon the chemistry of a crystalline material, including impurities it may contain, certain elements in this material may preferentially locate at the regions separating individual crystals, that is, crystal or grain boundaries. This may occur at high temperatures or for other reasons and is called grain boundary segregation.

Grain structure In polycrystalline materials, each of the crystals composing individual grains can have a different orientation or even composition relative to its neighbors. These structural differences are referred to as grain structure.

Graphite The hexagonal crystal structure for carbon. By comparison, carbon arranged in a special cubic crystalline structure characterizes diamond.

Graphitized phase A region in which carbon has assumed the hexagonal crystalline structure forming graphite.

Hardness testing Hardness is generally described as the reistance to penetration by a substance. If something is very hard it is very difficult to penetrate it. Penetration can be measured by the ability to scratch a surface or by the depth of a scratch, or the depth a penetrator will move into a substance for specific loads or forces applied. Testing can be performed by performing measurements of penetration and comparing the specific or relative magnitudes.

Hooke's law The law that stress σ applied to a solid material is proportional to the strain ϵ: $\sigma = E\epsilon$, where the constant of proportionality E is called the elastic modulus or Young's modulus.

Intergranular brittle fracture Brittle fracture that occurs at the interfaces or boundaries separating individual crystals or grains in a polycrystalline solid. This is usually an indication that the grain boundaries have been weakened as a result of segregation of some kind or of some related alteration in the structure.

Ion microscope A device that uses an ion beam to image magnified features of an object or substance. Ions ejected from or transmitted through the substance might be used to form the image.

Ion spectroscopy The separation of a beam of mixed ions into individual components arranged or separated according to their mass or some other partitioning parameter.

Lattice parameter A fundamental or unit dimension of a crystalline cell or crystalline unit cell. It is the distance between atoms characterizing the edges of the cell. It is also called the lattice constant. For cubic crystals it is the distance between the atoms at the corners of the cube, measured along an edge.

Liability An obligation or responsibility; anything for which a person is liable.

Liquid-phase chemical analysis An analysis in which the substance to be analyzed is a liquid. Solid substances must therefore be dissolved or reacted to be in a liquid form in order that its chemistry might be determined by liquid-phase chemical analysis.

Litigation The act or process of carrying on a civil action (or tort) or lawsuit.

Magnetic lens An arrangement of solenoids, electromagnets, or permanent magnets having axial symmetry, which creates magnetic fields that can be shaped to act upon charged particles and particle beams similar to the way a glass lens acts on light beams. In an electromagnetic lens, changes in the current in the associated lens coils can change the lens shape and thereby alter the convergence of beams of charged particles having uniform velocity.

Metallography The study and characterization of the structure of metals, alloys, and materials in general by various analytic methods employing light, electrons, ions, X-rays, and so on.

Microchemistry The chemistry of very small areas or volumes. For example, a large piece of a metal may contain a very minute impurity or precipitate. This very small inclusion will have a different composition and can be regarded in the context of the larger solid to have a unique microchemistry.

Microstructure As a general definition, microstructure can be any structure of an object or substance as revealed by a microscope at magnifications greater than about 10X.

Modulus of elasticity A measure of the incremental stress required to produce incremental strain. It is the ratio of stress divided by strain, sometimes called Young's modulus or the elastic modulus.

Multichannel analyzer A computer for storing signal information in specific locations corresponding to the signal amplitude or voltage. The system can actually sort the individual voltage pulses in the signal and add them in series to produce a measure of the component strength. Simultaneous display of this information in all channels produces a spectrum of the information.

Ordered alloys Combinations of elements forming metals can take place by having the individual atoms occupy random positions in the crystalline structure. Under certain conditions these same compositions can occur but the individual atoms occupy very specific positions in the lattice. This is an ordered structure, and such alloys are ordered alloys.

Phase diagram A graphic representation of the relationships among the solid, liquid, and vapor phases of a substance as a function of pressure, temperature, and composition. Two-dimensional diagrams usually represent phase transitions as a function of temperature and pressure or temperature and composition.

Phase equilibria The equilibrium relationships among solid, liquid, and vapor phases of different substances or the same substance under different conditions of temperature, pressure, and composition.

Photovoltaic device Any device that provides energy from sunlight. A solar cell is a common photovoltaic device, which is usually a semiconductor device that produces electricity from the sun's radiation.

Piezoelectric crystal Certain crystals, because of their structure and in particular their lack of symmetry, will create a small electrical voltage in proportion to a mechanical force exerted upon the crystal. This process can also work in reverse, using an electrical signal to create a mechanical response in the crystal. This phenomenon, the conversion of mechanical energy to electrical energy, and vice versa, will occur only along certain directions for such crystals.

Planck's constant A fundamental physical constant defined by the ratio of the energy of light or a photon to its frequency equal to 6.62620×10^{-34} joule-second. It is symbolized by h, and in quantum mechanics is also known as the quantum of action.

Poisson's ratio In mechanics or mechanical property evaluation the ratio of the transverse contracting strain to the elongation strain is called Poisson's ratio and is denoted by v. For a round bar pulled in tension it is the ratio of the reduction in the radius to the extension of the rod.

Polymorphic phase transformation A phase transformation that has different forms or structures. For example a solid, such as diamond, could transform to graphite or to another crystalline structure, all having different arrangements of carbon atoms.

Radiography The technique for producing a photographic image of an opaque substance by transmitting a beam of X-rays or other radiation through it and onto a photographic plate or other suitable imaging device. The resulting image results by contrast differences produced at less or more dense regions in the substance, including differences in thickness.

Raster scanning Controlling the intensity of the electron beam during sweep in a process similar to facsimile recording, in which characters are generated in segments by passing the beam through an appropriate aperture within a CRT. It is therefore a method for information or image display on a CRT screen.

Resolution or resolving power   A quantitative measure of the ability of an optical instrument (including a light microscope and an electron microscope) to produce separable images of two different points or other features on an object.

Safety factor In considering a working stress in a design or structure, the working stress S_w is obtained by dividing the yield σ_y or ultimate strength by a number called a factor of safety N_y: $S_w = \sigma_y/N_y$. The selection of the value for the safety factor calls for the designer's best judgment, so that neither unsafe nor uneconomical members are obtained in a working structure or system.

Scanning electron microscope A device for forming magnified images of the surface structure of materials by scanning a focused electron beam over a very small area of the surface and detecting backscattered or secondary electrons emitted from the scanned area. Emitted secondary electrons are detected and displayed according to the signal intensity on a cathode-ray tube (CRT) synchronously with the scan of the beam across the area. The system is analogous to a television studio in which the substance assumes the role of the scenery to be viewed and the detector acts like a TV camera, transmitting its signal to a CRT or image monitor.

Secondary electrons When a primary beam of electrons strikes a solid, electrons actually within the solid, such as conduction electrons not bound to particular atoms, are ejected. These are referred to as secondary electrons and notably differ from primary electrons by their comparatively smaller energies.

Secondary ion An ion or ions composing a material or solid that are expelled from the solid or material when it is bombarded by a primary ion beam. These component ions are charged atoms or molecules composing the solid, usually in or near a surface being bombarded. Different ions are measured by separating them into different regimes by passing them through a magnetic field.

Secondary-ion mass spectrum When a solid is bombarded by a primary beam of ions, the atoms or molecules composing the bombarded area can also be released and ionized. These ions represent the proportion of specific atoms composing the solid and can be separated into specific groups of ions according to their mass or charge-mass ratio. A display of these groups according to their relative or specific concentrations produces a spectrum of the solid composition.

Shear modulus Also called the modulus of elasticity in shear, it is a measure of a material's resistance to a shearing stress, equal to the shearing stress divided by the resultant angle of deformation. It is also called the coefficient of rigidity.

Solidification The change of a liquid (or a gas) into a solid.

Specialty steel A variety of steels (iron alloys) that have unique chemical compositions that allow them to be heat treated or otherwise fabricated to have unique properties or applications. Examples include types of stainless steels or very hard steels for tool applications.

Spectrometer An instrument by means of which the spectrum of wavelengths or energies of light or other beams, such as electrons or ions, can be accurately measured. An ion beam can in fact be separated into specific atomic components and the actual or relative magnitude of the components measured. See Ion Spectroscopy.

Speed of light The speed of propagation of light or electromagnetic waves representing light in a vacuum is a physical constant denoted by c and equal to 299,792.46 kilometers per second. It is the c in the famous Einstein equation $E = mc^2$, where E is the total system energy and m is its mass. It is also known as the electromagnetic constant.

Stiffness In the context of a deformable solid, stiffness is the ratio of a force required to produce some specific elastic displacement (or strain). An elastic displacement is one that will disappear or reverse when the force is removed. Consequently, a simple example would involve the force required to produce some unit compression or stretch of a spring. Stiff materials require larger forces for some unit compression or stretch.

Stoichiometry The chemistry of compounds or the proportion of elements composing compounds. Also, the treatment of combining elements or compounds involved in chemical reactions.

Strain Denoted by epsilon (ϵ), strain is a change in a dimension of an object compared with the original dimension. For example, the strain involved in stretching a copper wire will be the change in its length after an applied force divided by its original length.

Strain rate The displacement or incremental displacement of a body with some incremental time; strain per unit time.

Strain state (stress state) The state of straining or stressing a body is related to the degrees of freedom involved in the displacement of elements of a body in a three-dimensional system; each of the three axes (x, y, and z) is devoted to a particular principal stress (σ_1, σ_2, and σ_3). This stress or strain state or system is called uniaxial if only one principal stress (or strain) is active (σ_1), biaxial if two are involved (σ_1 and σ_2), and triaxial if all three are involved.

Stress A force acting across a unit area in a solid is called a stress, denoted by sigma (σ) or S. This stress tends to be induced by external forces acting on a unit area. In stretching a copper wire, for example, the stress is the load or force applied in stretching the wire divided by the cross-sectional area [equal to pi (π) times the square of the radius]. This stress acts normal to the cross-sectional areas and is called a tensile stress.

Stress corrosion cracking Corrosive attack creates notches (many times at grain boundaries or crystal boundaries in polycrystalline metals and alloys), which under an applied stress tend to concentrate the stress and cause cracks to form at the root of the notch.

Structure transformations When iron having a body-centered cubic crystal structure is induced to alter its structure to a face-centered cubic structure, this is regarded as a structural transformation. Many solids undergo such a transformation. Under certain conditions, graphite, which is the hexagonal crystalline structure of carbon, can transform to the cubic crystalline structure, the most common of which would be diamond.

Submicroscopic Very small features that are difficult or impossible to see with a light microscope can be regarded as below the resolution of the microscope and therefore submicroscopic. Such detail may require an electron microscope or other device to view it. Objects smaller than about 0.2 micrometers are normally not observable by light microscopes, and therefore submicroscopic details might be regarded as having features below this value.

Tool steel Any variety of steel (iron alloy) capable of being hardened sufficiently so as to be suitable for cutting tool or machine tool applications.

Tort A wrongful act, injury, or damage not involving a breach of contract for which a civil action can be brought.

Toughness The ability of a material to absorb energy by plastic deformation. Plastic deformation is permanent deformation, so a very tough material would be one that could be deformed by a large force and remain intact. Neither very soft nor very brittle materials possess the property of toughness.

Transmission electron microscope A device for forming magnified images of very thin materials or substances by transmitting electrons in a focused beam through them. The images are formed by focusing the transmitted electrons by either electrostatic (electric) or magnetic lenses and projecting them onto a fluorescent screen.

Triaxial loading Three principal stresses (corresponding to directional axes x, y, and z) operate on any point in a body; a hydrostatic stress.

Ultrasonic waves Sound waves that normally have a frequency above about 20 kilohertz (or 20,000 cycles per second).

Uniaxial loading The application of a load, force, or resolved stress in one direction to deform a solid. For example, pulling a wire involves a uniaxial, tensile load.

Unit cell The smallest recognizable, regular crystal unit characterized by the arrangement of atoms. Crystals are built up of repetitions of the unit cell. The unit cell can be visualized as a space structure; there are 14 unique such structures, called Bravais lattices. Ideally, therefore, there are 14 unique unit cells that can characterize all crystals.

Viscosity Viscosity is generally considered a quantitative measure of the resistance of a substance or material to continuous shearing. A high viscosity therefore denotes a high resistance to shearing. Many substances have lower viscosity at higher temperature. Molasses is a good example. Substances that do not flow easily are obviously highly viscous.

Wavelength Fundamentally defined as the distance between two points having the same phase in two consecutive cycles of a periodic wave, along a line in the direction of propagation. This distance is usually denoted by λ and can be illustrated by the distance between wave maxima as shown in the illustrations. Wavelength is found by dividing the speed of light by the frequency.

Weld porosity Gas bubbles and voids trapped in a weld zone constitute a kind of porosity commonly referred to as weld porosity. This porosity can be measured as the ratio of void volume to actual volume and cannot exceed certain standards for particular types of welds.

Work hardening An increase in the hardness of a material as it is worked or plastically deformed. For example, consider that a copper pipe can be easily bent the first time but it will require more force to bend it back, and slightly more force to bend it again, and so on. Thus, it becomes harder as it is worked.

X-ray diffraction The scattering of X-rays or a beam of X-rays from matter. In crystals the scattering is systematic. The scattered X-ray intensity can vary because of interference effects, which can also be systematic.

X-ray shadowgraph A photographic image produced by X-rays. Voids and impurities or other irregularities in solid materials will appear in contrast as shadows or outlines of these features.

Bibliography

FUNDAMENTALS OF MATERIALS SCIENCE AND
ENGINEERING

Azaroff, L. V., *Introduction to Solids*, Krieger, Inc., New York, 1976.

Billmeyer, Jr., F. W., *Textbook of Polymer Science*, 2nd edition, Wiley, New York, 1971.

Brick, R. M., Pease, A. W., and Gordon, R. B., *Structure and Properties of Engineering Materials*, McGraw-Hill, New York, 1977.

Eisenstadt, M. M., *Introduction to Mechanical Properties of Materials*, Macmillan Publishing Co., New York, 1971.

Guy, A. G., *Introduction to Materials Science*, McGraw-Hill, New York, 1972.

Kingery, W. D., *Introduction to Ceramics*, J. Wiley & Sons, New York, 1960.

Lewis, T. J., and Secker, P. E., *Science of Materials*, Reinhold Publishing Corp., New York, 1965.

Murr, L. E., *Solid-State Electronics*, Marcel Dekker, New York, 1978.

Rosenberg, H. M., *The Solid State: An Introduction to the Physics of Crystals for Students of Physics, Materials Science, and Engineering*, Clarendon Press, Oxford, 1978.

Suh, N. P., and Turner, A. P. C., *Elements of the Mechanical Behavior of Solids*, McGraw-Hill, New York, 1975.

CRYSTAL IMPERFECTIONS, DEFECTS IN MATERIALS, AND PHASE EQUILIBRIA

Barrett, C. S., and Massalski, T. B., *Structure of Metals*, 3rd edition, McGraw-Hill, New York, 1966.

Diehl, J., *Vacancies and Interstitials in Metals*, North-Holland, Amsterdam, 1970.

Gordon, P., *Principles of Phase Diagrams in Materials Systems*, McGraw-Hill, New York, 1968.

Henderson, B., and Hughes, A. E., *Defects and Their Structure in Non-metallic Solids*, Plenum Publishing Corp., New York, 1976.

Kelly, A., and Groves, G. W., *Crystallography and Crystal Defects*, Addison-Wesley Publishing Co., Reading, Massachusetts, 1975.

Prince, A., *Alloy Phase Equilibria*, American Elsevier Publishing Co., New York, 1966.

Read, W. T., *Dislocations in Crystals*, McGraw-Hill, New York, 1953.

MATERIALIZATION CHARACTERIZATION USING LIGHT, SOUND, X-RAY, ELECTRONS, AND IONS

Amelinckx, S., Gevers, R., Remaut, G., and VanLanduyt, J. (eds), *Modern Diffraction and Imaging Techniques in Materials Science*, North-Holland American Elsevier, New York, 1970.

Clark, G. C., *Encyclopedia of Microscopy*, Reinhold Publishing Corp., New York, 1961.

Cohen, J. B., *Diffraction Methods in Materials Science*, The Macmillan Company, New York, 1960.

Czanderna, A. (ed.), *Methods of Surface Analysis*, Elsevier, Amsterdam, 1975.

Gifkins, R. C., *Optical Microscopy of Metals*, Elsevier, New York, 1970.

Kane, P. F., and Larrabee, G. B. (eds.), *Characterization of Solid Surfaces*, Plenum, New York, 1974.

Muller, E. W. and Tsong, T. T., *Field-ion Microscopy: Principles and Applications*, Elsevier, New York, 1969.

Murr, L. E., *Electron Optical Applications in Materials Science*, McGraw-Hill, New York, 1970.

Murr, L. E., *Electron and Ion Microscopy and Microanalysis: Principles and Applications*, Marcel Dekker, Inc., New York, 1982.

Rostoker, W., and Dvorak, J. R., *Interpretation of Metallographic Structures*, Academic Press, New York, 1965.

Smallman, R. E., and Ashbee, K. H. G., *Modern Metallography*, Pergamon Press, Oxford, 1966.

Taylor, A., *X-ray Metallography*, J. Wiley, New York, 1961.

MATERIALS TESTING, FAILURE ANALYSIS, AND FRACTOGRAPHY

Avery, D. G., and Findley, W. N., Quasistatic Mechanical Testing, in *Techniques of Metals Research*, Vol. 5, Part I, R. F. Bunshah (ed), Wiley-Interscience, New York, 1971.

Davis, H. E., Troxell, G. E., and Wiskocil, C. T., *The Testing and Inspection of Engineering Materials*, McGraw-Hill, New York, 1941.

Fractography and Atlas of Fractographs, Metals Handbook, Vol. 9, 8th edition, ASM, Metals Park, Ohio, 1974.

Hertzberg, R. W., *Deformation and Fracture Mechanics of Engineering Materials*, Wiley, New York, 1976.

McCall, J. L., and French, P. M. (eds.), *Metallography in Failure Analysis*, Plenum Publishing Corp., New York, 1981.

Meyers, M. A., and Chawla, K. K., *Mechanical Metallurgy: Principles and Applications*, Prentice-Hall, Inc., Englewood Cliffs, New Jersey, 1984.

Petty, E. R., Hardness Testing, in *Techniques of Metals Research*, Vol. 5, Part 2, R. F. Bunshah (ed.), Wiley-Interscience, New York, 1971, p. 157.

Westbrook, J. H., and Conrad, H. (eds.), *The Science of Hardness Testing and Its Research Applications*, ASM, Metals Park, Ohio, 1973.

FAILURE PREVENTION IN MATERIALS SYSTEMS: ENGINEERING DESIGN AND SAFETY FACTORS

Collins, J. A., *Failure of Materials in Mechanical Design: Analysis, Prediction, Prevention*, Wiley-Interscience, New York, 1981.

Dieter, George, *Engineering Design: A Materials and Processing Approach*, McGraw-Hill Book Co., New York, 1983.

Faupel, J. F., and Fisher, Franklin E., *Engineering Design: A Synthesis of Stress Analysis and Materials Engineering*, Wiley-Interscience, New York, 1981.

Karajian, G. M. (ed.), *Failure Prevention and Reliability*, American Society of Mechanical Engineers, 1983.

Murr, L. E., *Interfacial Phenomena in Metals and Alloys*, Addison-Wesley, Reading, Massachusetts, 1975.

Perry, Robert H., *Engineering Manual: A Practical Reference of Design*, McGraw-Hill Book Co., New York, 1976.

RESPONSIBILITY, LIABILITY, AND LITIGATION

Green, Leon, *Litigation Process in Tort Law*, 2nd edition, The Michie Co., Charlottesville, Virginia, 1977.

Lieberman, Jethro, *Litigious Society*, Basic Books, Inc., New York, 1983.

Thorpe, James F., and Middendorf, William H., *What Every Engineer Should Know About Product Liability*, Marcel Dekker, Inc., New York, 1979.

Weinstein, A. S., et al., *Products Liability and the Reasonably Safe Product: A Guide for Management, Design, and Marketing*, John Wiley & Sons, New York, 1978.

Witherell, Charles, *How to Avoid Products Liability, Lawsuits and Damages*, Noyes Publications, Park Ridge, New Jersey, 1985.

Index